STATE OF WASHINGTON

DEPARTMENT OF NATURAL RESOURCES

BERT L. COLE, Commissioner of Public Lands
DON LEE FRASER, Supervisor

DIVISION OF GEOLOGY AND EARTH RESOURCES

VAUGHN E. LIVINGSTON, JR., State Geologist

BULLETIN NO. 69

SILVER OCCURRENCES

OF

WASHINGTON

By

WAYNE S. MOEN

1976

For sale by Department of Natural Resources, Olympia, Washington
Price $4.00

CONTENTS

PART I—HISTORY, PRODUCTION, GEOLOGY, AND MINERALOGY OF SILVER IN WASHINGTON

CONTENTS

CONTENTS

PART III—SILVER OCCURRENCES OF WESTERN WASHINGTON

CONTENTS

ILLUSTRATIONS

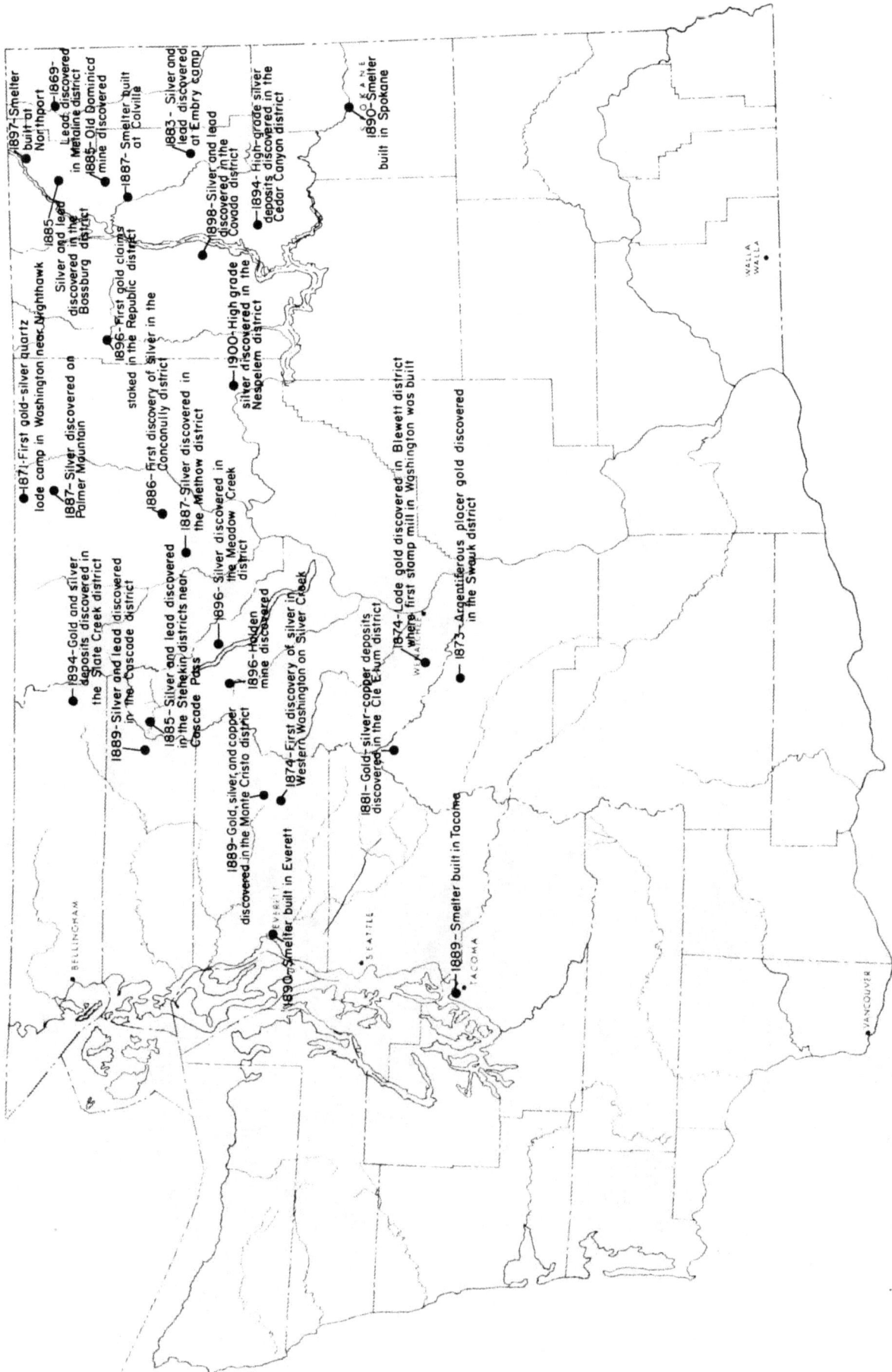

1897-Smelter built at Northport

1869-Lead discovered in Metaline district

1885-Old Dominica mine discovered

1887-Smelter built at Colville

1883 - Silver and lead discovered at Embry camp

1885-Silver and lead discovered in the Bossburg district

1898-Silver and lead discovered in the Covada district

1894-High-grade silver deposits discovered in the Cedar Canyon district

1890-Smelter built in Spokane

1896-First gold claims staked in the Republic district

1900-High grade silver discovered in the Nespelem district

1871-First gold-silver quartz lode camp in Washington near Nighthawk

1887-Silver discovered on Palmer Mountain

1886-First discovery of silver in the Conconully district

1887-Silver discovered in the Methow district

1894-Gold and silver deposits discovered in the Slate Creek district

1889-Silver and lead discovered in the Cascade district

1885-Silver and lead discovered in the Stehekin districts near Cascade Pass

1896-Silver discovered in the Meadow Creek district

1896-Holden mine discovered

1874-First discovery of silver in Western Washington on Silver Creek

1874-Lode gold discovered in Blewett district where first stamp mill in Washington was built

1873-Argentiferous placer gold discovered in the Swauk district

1889-Gold, silver, and copper discovered in the Monte Cristo district

1881-Gold-silver-copper deposits discovered in the Cle Elum district

1890-Smelter built in Everett

1889-Smelter built in Tacoma

Historical events relating to early silver mining in Washington

SILVER OCCURRENCES of WASHINGTON

by

Wayne S. Moen

ABSTRACT

Silver was one of the first metals utilized by man, for as early as 2500 to 3000 B.C. the inhabitants of Asia Minor used silver for ornaments and utensils. The Romans were the first to use the metal as the basis of a monetary system, and for hundreds of years silver was used in coins, as well as in jewelry and sterling ware. Currently (1975), the photographic and electronic industries are the largest consumers of silver in the United States. Since 1942, the industrial demand for silver has exceeded our domestic production, and each year the silver deficit increases. In 1973, mines in the United States produced only 37 million ounces of silver, while the net industrial consumption was 193 million ounces. The increasing demand for silver has resulted in higher silver prices. In 1960, silver sold for around 91 cents per troy ounce; in February 1974 it reached a high of $6.70 per ounce; in August 1975, it sold for around $4.50 per ounce. Shortages of silver combined with increasing demands in the future will undoubtedly force silver prices even higher.

Washington never has been a major silver-producing state, but it has been a consistent producer of silver since 1860. From 1860 through 1972, mines in the state produced 22.8 million ounces of silver valued at $20.3 million. The bulk of this silver was a byproduct of gold, copper, lead, and zinc mining operations in Ferry, Chelan, Stevens, and Pend Oreille Counties. In 1972, Washington produced 220,782 ounces of silver as a byproduct of gold mining operations at Republic in Ferry County and lead-zinc mining operations at Metaline Falls in Pend Oreille County.

Silver has been reported at over 1,500 mines and prospects in 26 of the state's 39 counties. However, only at 268 properties does it appear to be present in significant amounts. Counties in which the principal silver deposits are located include Pend Oreille, Stevens, Ferry, Okanogan, Chelan, Kittitas, King, Snohomish, Skagit, and Whatcom.

The silver deposits of Washington occur mainly in narrow quartz fissure veins in a variety of host rocks, which range in age from Paleozoic to Tertiary. However, most deposits appear to be of Cretaceous and early Tertiary age, and are related to granitic intrusions. Ore minerals common to the veins include pyrite, pyrrhotite, galena, chalcopyrite, and sphalerite. The less common ore minerals are tetrahedrite, bornite, stibnite, silver sulfides, native silver, and gold. In many deposits, silver is commonly carried by the galena and tetrahedrite, while in several of the richest deposits mined in the past silver minerals such as pyrargyrite, cerargyrite, argentite, and native silver were predominant.

The silver content of the veins varies considerably. Lead-zinc ore from Stevens and Pend Oreille Counties averages only around 0.02 ounce per ton. Copper ore from Stevens County averaged around 4 ounces per ton in silver, as did the ore from several gold mines in Ferry County. Several silver mines in Stevens, Okanogan, and Snohomish Counties shipped ore that contained several hundred ounces per ton in silver, whereas select high-grade ore yielded up to 1,000 ounces per ton. In deposits that contained mainly silver, the average silver content of ore mined in the past was round 60 ounces per ton. However, the average silver content of the principal silver mines and prospects in Washington is only

around 6 ounces per ton.

It is doubtful that any silver mines or prospects in the state will develop into major silver producers like silver mines of the Coeur d'Alene mining district of Idaho. Several properties might develop into small producing mines, and exploration might turn up new ore bodies. However, as in the past, the bulk of the silver produced in Washington in the future will be produced as byproducts of gold, copper, lead, and zinc mining operations.

GENERAL DISCUSSION

PROPERTIES, USES, AND CONSUMPTION

Silver, like gold, was one of the first metals cherished by man. As early as 2500 to 3000 B.C., it was fashioned into ornaments and utensils, and for over 2,000 years was the basis of monetary systems throughout the world.

Silver is the whitest of all metals, has a perfect metallic luster, and except for gold is the most malleable and ductile of all metals. Because of its comparative scarcity, brilliant white color, and resistance to atmospheric oxidation, it has been used for centuries in coins, jewelery, ornaments, and silverware. It is one of the best metals for conducting electricity and it is widely used in the manufacture of electrical and electronic products. Because of its unusual chemical properties, it is used extensively by the photographic industry. In 1973, 26 percent of the silver consumed in the United States was used in photographic materials; 22 percent was used in electrical and electronic products; 20 percent was used in sterling ware; 16 percent was used in electroplated and brazing wares; and 8 percent was

used in medical and dental products, catalysts, bearings, and jewelry. Because of silver shortages throughout the world, it has not been used in coins since the late 1960's. However, many commemorative coins are still minted of silver.

Prior to 1942, production of silver in the United States exceeded industrial consumption. However, beginning in 1942, the industrial demand for silver increased rapidly, and by 1974 the net industrial consumption was 178 million ounces, while the mines of the country produced only 34 million ounces (fig. 1). Up until 1971, the silver deficit was balanced by the withdrawal of silver from the U.S. Treasury, as well as by sales of hoarded and reclaimed silver. However, the depletion of Treasury silver has forced silver users to depend upon newly mined silver, reclaimed silver, sale of private silver, and imports of silver from foreign countries. Faced with a silver deficit in 1974, the United States imported 134 million ounces of silver from 16 foreign countries, with the largest imports coming from Canada, Mexico, and Peru.

Silver is a comparatively rare metal, and the average rock is estimated to contain only 0.07 parts

AVERAGE U.S. SILVER PRICES

(1860-1974)

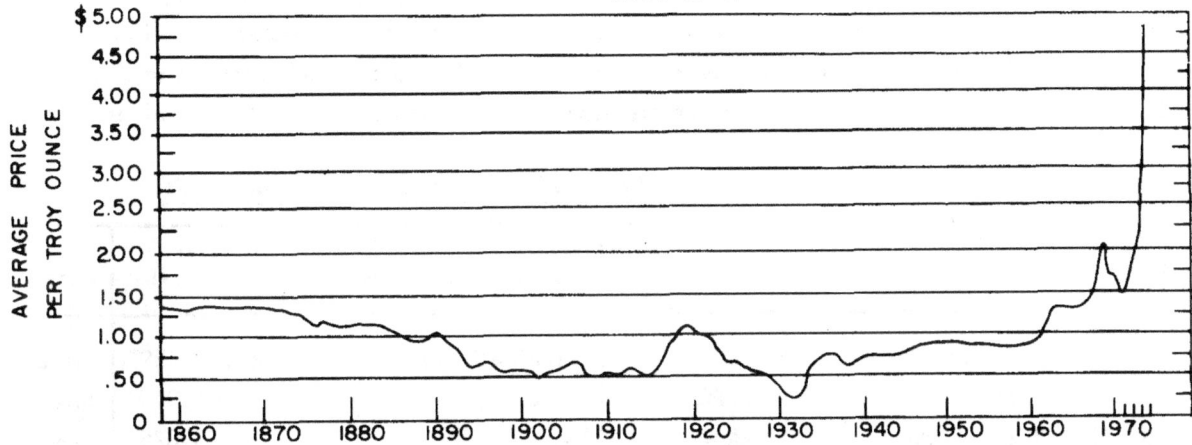

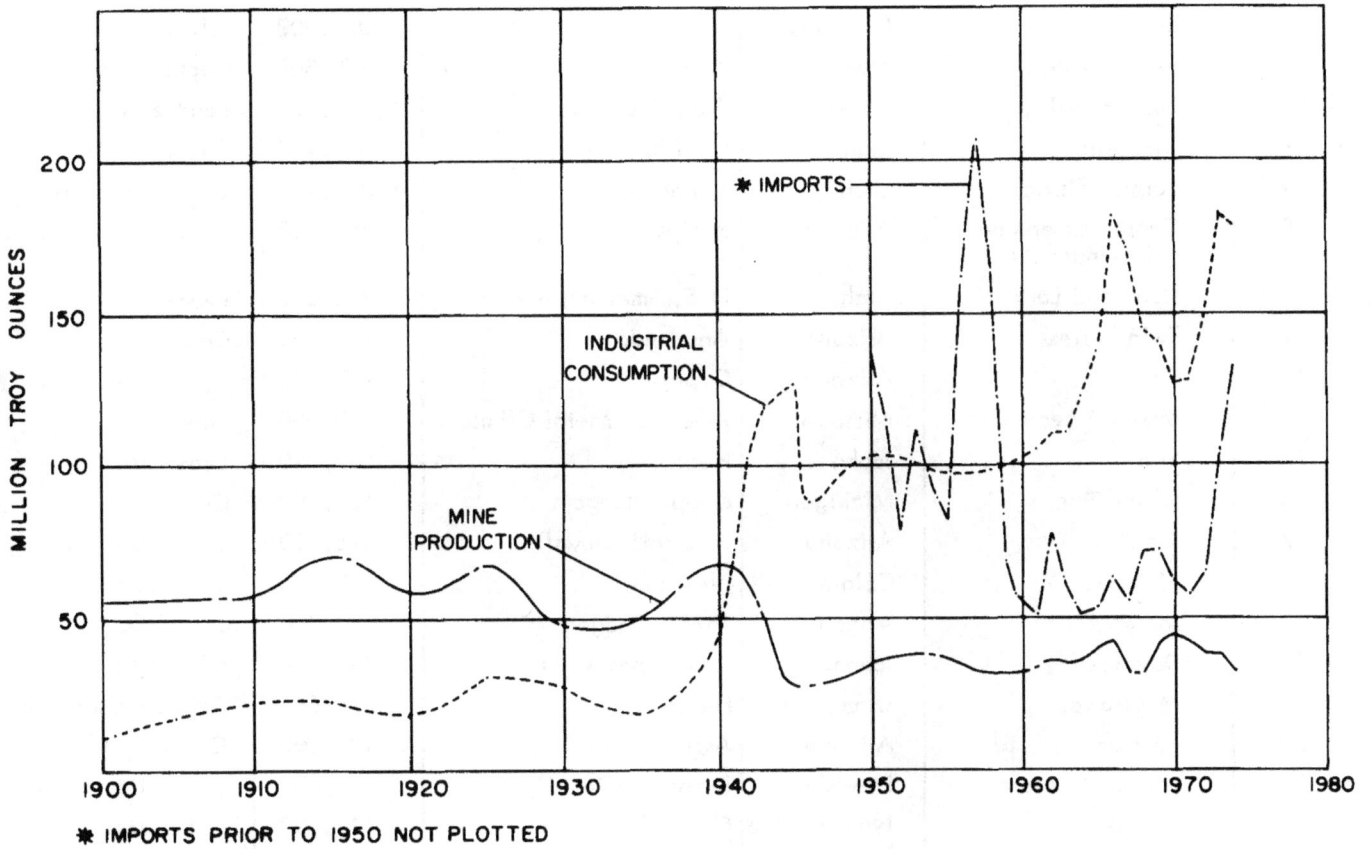

FIGURE 1.—Mine production, industrial consumption, and silver prices in the United States, 1900-1974.

per million (0.002 troy ounces per ton) of the metal. Mines in the United States that are classed as true silver mines are scare. Only four silver mines produced over 1 million troy ounces of silver in 1971, and three of the mines were in the Coeur d'Alene mining district of Idaho (table 1). Silver ore from three of the silver mines averaged up to 27 ounces of silver per ton, which is 13,500 times that of the average rock. However, the bulk of the silver mined in the United States comes from base metal mines, the ore of which contains as little as 0.06 ounces of silver per ton. Although silver is only a byproduct

TABLE 1.—Major silver producers in the United States, 1971

Rank	Mine	State	Operating Company	Silver (troy ozs)	Class of ore
1	Sunshine	Idaho	Sunshine Mining	7,043,629	Silver
2	Galena	Idaho	Asarco	3,901,000	Silver
3	Lucky Friday	Idaho	Hecla	3,343,075	Lead
4	Utah Copper	Utah	Kennecott	2,291,969	Copper-gold-silver
5	Bulldog	Colorado	Homestake	2,018,300	Silver
6	Berkeley Pit	Montana	Anaconda	1,970,301	Copper
7	Bunker Hill	Idaho	Gulf Resources	1,715,022	Lead-zinc
8	Crescent	Idaho	Gulf Resources	1,663,417	Silver
9	Burgin (Tintic)	Utah	Kennecott	1,029,975	Lead, lead-zinc
10	Copper Queen and Lavender Pit	Arizona	Phelps Dodge	867,000	......Do......
11	U.S. and Lark	Utah	U.S. Smelting & Refining	858,000	Lead-zinc
12	Twin Buttes	Arizona	Anaconda	796,236	Copper
13	Pima	Arizona	Cyprus	778,000	......Do......
14	Missouri Lead	Missouri	American Metal Climax	700,000	Lead
15	Star	Idaho	Hecla	647,340	Lead-zinc
16	White Pine	Michigan	Copper Range	634,780	Copper
17	Sierrita	Arizona	Pennzoil (Duval)	616,000	......Do......
18	Idarado	Colorado	Newmont	602,038	......Do......
19	Morenci	Arizona	Phelps Dodge	595,000	......Do......
20	Dayrock	Idaho	Day Mines	593,184	Lead-zinc
21	Mayflower	Utah	Hecla	576,663	Copper-lead-zinc
22	Mission	Arizona	Asarco	489,000	Copper
23	Battle Mountain	Nevada	Pennzoil (Duval)	462,000	......Do......
24	Tyrone	New Mexico	Phelps Dodge	446,000	......Do......
25	Butte Hill	Montana	Anaconda	429,416	......Do......
26	Mineral Park	Arizona	Pennzoil (Duval)	427,000	......Do......
	Total			35,494,345	

TABLE 2.—Production of silver in the United States, by class of ore, in 1971[1]

Class of ore	Ore (tons)	Silver (troy ozs)	Average grade (troy ozs/ton)	Percentage of total ore production
Silver	673,116	15,044,825	22.35	36.0
Copper	217,245,002	13,174,914	0.060	32.0
Copper-zinc and copper-lead-zinc..	5,115,314	6,826,542	1.334	17.0
Lead	8,925,109	5,825,447	0.652	14.0
Gold	1,882,638	405,443	0.215	1.0
Zinc	102,299	24,910	0.243	less than 0.5
Gold-silver	166,660	122,880	0.737	less than 0.5
Old tailings	100,809	138,076	1.369	less than 0.5

[1] From U.S. Bureau of Mines Minerals Yearbook, 1971, v. 1, p. 1081-1082.

TABLE 3.—Mine production by states of recoverable silver in the United States in 1973[1]

Rank	State	Silver (troy ozs)
15	Alaska	260
2	Arizona	6,974,010
11	California	67,492
4	Colorado	3,762,920
1	Idaho	13,575,596
8	Michigan	850,273
6	Missouri	2,122,197
3	Montana	4,128,149
9	Nevada	562,415
7	New Mexico	1,118,611
12	New York	54,010
14	Oregon	1,036
10	South Dakota	71,937
13	Tennessee	33,310
5	Utah	3,721,882
	Other states	188,675
	Total	37,232,773

[1] Preliminary figures from U.S. Bureau of Mines.

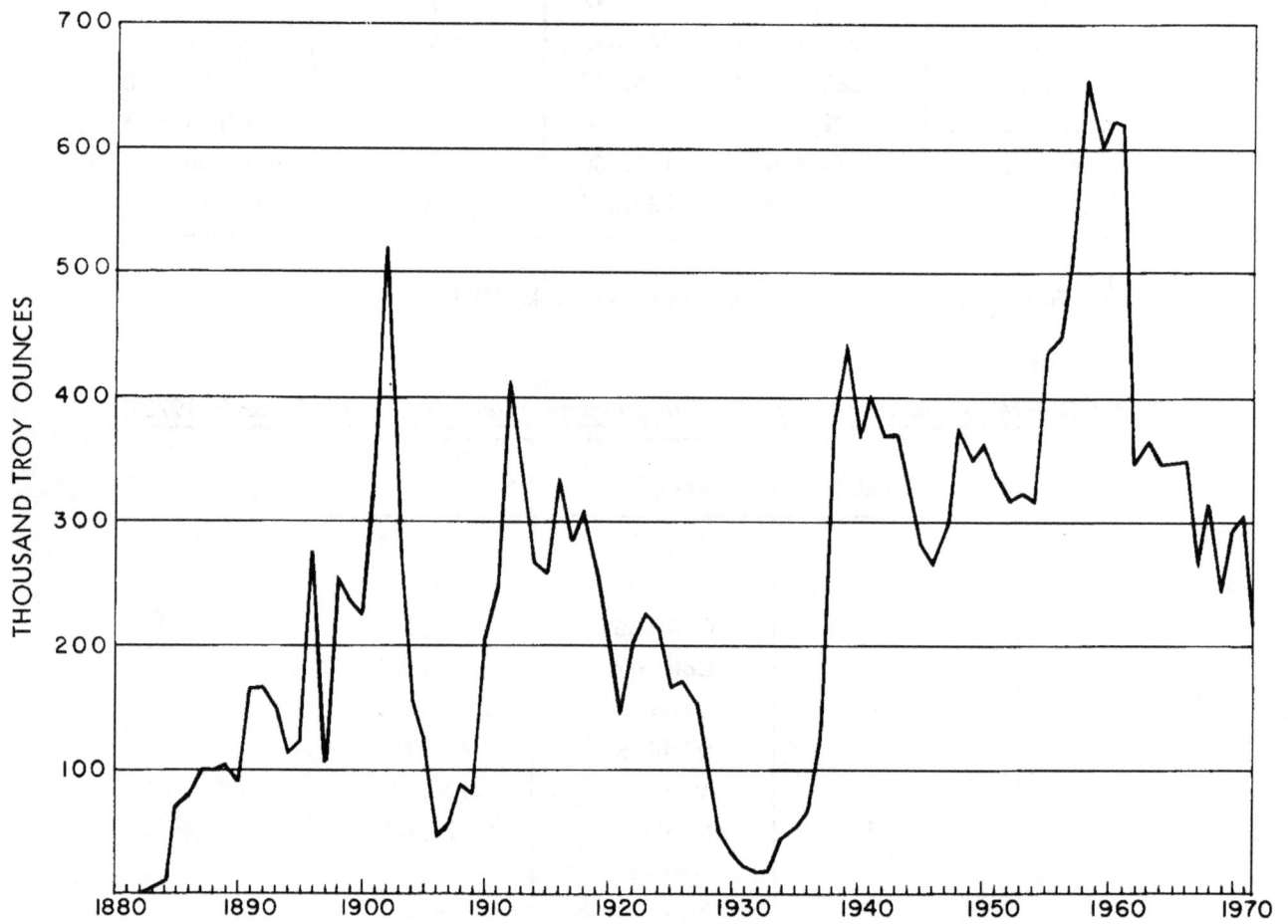

FIGURE 2.—Production of silver in Washington, 1882-1970.

of the base metal mines, they produced 63 percent of the silver mined in the United States in 1971. Table 1 lists the 26 largest silver mines in the United States; the combined production of these mines accounted for 85 percent of the silver produced in 1971. Table 2 lists the production of silver during 1971 according to ore types.

A breakdown of silver (by states) produced in the United States during 1973, is shown in table 3. The largest silver-producing states were Idaho, Arizona, Montana, Colorado, Utah, and Missouri. Around 92 percent of the 37 million ounces of silver produced during 1973 came from these states. Washington's silver production is not listed because the bulk of the silver came from the Knob Hill gold mine, and publishing the figure would disclose individual company confidential data.

SILVER PRICES

Over the past 100 years the price of silver has been as low as 28 cents per troy ounce in 1932 to as high as $6.70 per ounce on February 26, 1974. Between 1860 and 1960, silver averaged 90 cents per ounce; however, following cessation of Treasury sales in November 1961, there has been a general rise in the price of silver. Silver prices are subjected to daily fluctuations, and prices are published daily in the market section of many newspapers. During the first nine months of 1975, silver averaged around $4.50 per ounce with a low of $3.94 in January and a high of $5.10 in August. Because silver consumption is expected to increase more rapidly than mine production, higher silver prices can be expected in the future.

WASHINGTON'S SILVER PRODUCTION

Washington has been a consistent, but minor, producer of silver since 1860. The state is credited with the production of small amounts of silver prior to 1885; however, this silver came from the refining of gold bullion. From 1860 through 1972, mines of the state produced 22.8 million ounces of silver valued at $20.3 million. The yearly production was as little as 9,683 ounces in 1894, to as much as 320 ounces in 1876 (see Appendix C). During the past 10 years, the state has produced an average of 300,000 ounces yearly. Prior to 1885, the production of gold in terms of troy ounces exceeded the production of silver. However, from 1885 to 1975, with the exception of 1894, ounces of silver produced has always exceeded ounces of gold produced.

In 1972, the production of silver in Washington amounted to 220,782 ounces, the lowest it had been since 1937. Whereas during some years as many as 15 base metal and precious metal mines produced silver, in 1975, silver was produced only as a byproduct of gold mining operations at the Knob Hill mine at Republic, and as a byproduct of lead-zinc mining operations at the Pend Oreille mine near Metaline Falls.

Production of silver from lode mines in Washington began in 1874, when the first lode gold mines were placed into production; however, silver at that time was secondary to gold. In 1885, silver began to be produced from high-grade lead-silver ores, and by 1900, the lead-silver mines were producing over one-half of the silver mined in the state. In 1916, the silver came mainly from copper ores, and by 1943, silver was produced at the state's lead-zinc, and copper mines. However, at this time the production of silver from lode gold-silver mines exceeded the silver production from base metal mines. By 1957, the production of silver as a byproduct of the base metal mines had declined to less than 10 percent of the state's total silver production. Currently (1975), 97 percent of the silver produced in Washington comes from gold ores and 3 percent comes

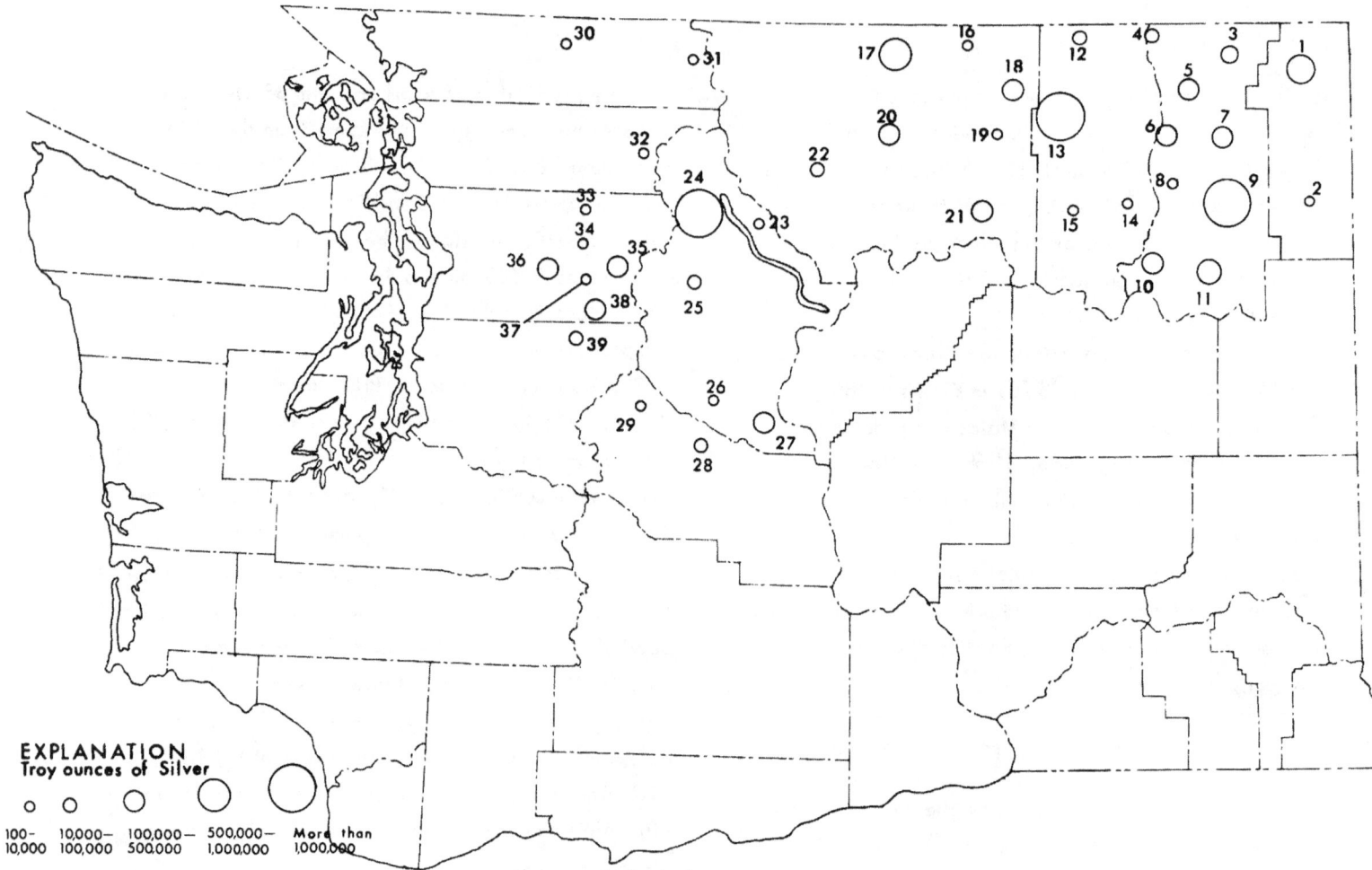

EXPLANATION
Troy ounces of Silver

| 100-10,000 | 10,000-100,000 | 100,000-500,000 | 500,000-1,000,000 | More than 1,000,000 |

FIGURE 3.—Total silver production by district, 1866-1973.

MINING DISTRICTS

Pend Oreille County

1 – Metaline
2 – Newport

Stevens County

3 – Northport
4 – Orient
5 – Bossburg
6 – Kettle Falls
7 – Colville
8 – Summit
9 – Chewelah
10 – Deer Trail
11 – Springdale

Ferry County

12 – Danville
13 – Republic
14 – Covada
15 – Keller

Okanogan County

16 – Myers Creek
17 – Loomis
18 – Sheridan
19 – Park City
20 – Conconully
21 – Nespelem
22 – Twisp

Chelan County

23 – Meadow Creek
24 – Railroad Creek
25 – Chiwawa
26 – Blewett
27 – Wenatchee

Kittitas County

28 – Swauk
29 – Cle Elum

Whatcom County

30 – Mount Baker
31 – Slate Creek

Skagit County

32 – Cascade-
      Thunder Creek

Snohomish County

33 – Darrington
34 – Silverton
35 – Sultan
36 – Monte Cristo
37 – Silver Creek
38 – Index

King County

39 – Miller River

from iead-zinc ores. Sources of silver in Washington from 1900 to 1972, based on ore types, is shown in table 4.

Ferry County is Washington's leading silver-producing county. Since 1904, it has produced around 7 million ounces of silver. Stevens and Chelan

TABLE 4.—Sources of Washington silver by classes of ore, 1900-1972

Class of ore	YEAR								
	1900	1910	1920	1930	1940	1950	1960	1970	1972
	Percentage								
Siliceous gold-silver ores	50	95	23	5	54	50	96	93	97
Copper ores	2	3	74	84	43	1	1	..	..
Lead ores	48	2	..	..	..	13	..	..	..
Lead-zinc ores	..	..	3	11	3	..	3	7	3
Zinc, zinc-lead, zinc-copper, zinc-lead, copper	..	..	..	..	..	37	..	..	..
	Tons								
Ore mined (all classes)	..	59,209	85,318	45,456	1,124,564	1,279,595	1,070,000	..	283,514
	Ounces per ton								
Average grade	..	3.47	2.34	0.72	0.39	0.85	0.58	..	0.78

TABLE 5.—Major silver-producing counties in Washington, 1904-1969

Rank	County	Silver (troy ounces)	Value (dollars)
1	Ferry[1]	5,917,851	$3,651,472
2	Stevens	3,378,763	2,257,584
3	Chelan	2,822,225	2,364,048
4	Okanogan	860,159	678,843
5	Pend Oreille	708,330	614,617
6	Snohomish	249,005	169,697
7	Whatcom	57,837	35,214
8	King	34,503	24,292

[1] Ferry County production from 1904 through 1956; confidential data after 1956 not disclosed.

counties are the only other counties having total silver productions that exceed 1 million ounces. Major
silver-producing counties of Washington are listed in
table 5. Production is from 1904 through 1969; prior
to 1904, silver production in the state was not broken
down by county.

OUTLOOK FOR SILVER
IN WASHINGTON

Although silver is widespread throughout the
Northern Cascades and the Okanogan Highlands of
northeastern Washington, very few silver deposits
appear to be of economic value. At most deposits,
silver is associated with other metals such as gold,
copper, lead, and zinc; only at a few deposits is
silver the predominant metal. As in the past, silver
will continue to be chiefly a byproduct of gold,
copper, lead, and zinc mining operations; however,
because of increasing silver prices, several mines
classed as silver mines could become important silver
producers. These mines are as follows: (1) United
Copper in the Chewelah district of Stevens County.
(2) Deer Trail and Queen-Seal mines in the Deer
Trail district of Stevens County. (3) Fourth of July,
Arlington, Last Chance, and First Thought mines in
the Conconully district of Okanogan County. (4)
Ruby and Ivanhoe mines in the Loomis district of
Okanogan County. (5) Apex and Cleopatra mines
in the Miller River district of King County. (6) "45"
mine in the Sultan district of Snohomish County.
(7) Great Excelsior mine in the Mount Baker district
of Whatcom County.

The writer does not wish to imply that these
mines are the only mines in Washington capable of
becoming silver producers. Other mines and prospects
may be equally as important, but sufficient data are
not available to properly evaluate them. Any increases in the price of silver will stimulate exploration for silver at silver districts covered in this report.
As in the past, Washington should continue to be a
minor but consistent producer of silver.

SCOPE AND PURPOSE OF REPORT

This bulletin represents a compilation of information on silver occurrences in Washington. It
contains data presented in a large number of published
and unpublished reports that are listed in the reference section. The majority of the silver occurrences
listed in this bulletin were obtained from "Inventory
of Washington Minerals, Part II, Metallic Minerals"
(Huntting, 1956). Other than verifying the location
of many occurrences, no new field work was done in
the preparation of this report.

The bulletin is divided in three parts: Part I
discusses Washington's silver industry, silver-mining
history, production, geology and mineralogy of silver
deposits, and distribution of silver in Washington;
Part II and III describe the principal silver-mining
areas of the state and give pertinent information for
individual mines and prospects. Part II covers the
Okanogan Highlands physiographic province of northeastern Washington, whereas Part III covers the Cascade Mountains province of western Washington.

The purpose of this bulletin is to place under
one cover the mines and prospects containing silver
so information is readily available to those who may
be interested in the state's silver occurrences, whether
it be academic, historic, or economic. Other purposes are to update data on individual properties and
to correct errors such as locations, production figures,
and geological data.

PART I

HISTORY, PRODUCTION, GEOLOGY, and MINERALOGY of SILVER in WASHINGTON

SILVER MINING HISTORY

EASTERN WASHINGTON, 1855-1900

DISCOVERY

One cannot discuss the history of silver mining in Washington without discussing metal mining in
general, because the bulk of the silver produced in
the state has come from gold, copper, lead, and zinc
mines. Also, the discovery and mining of placer
gold is significant because the first silver produced

Gold panner

in Washington came from the refining of placer gold,
some of which contained as much as 25 percent silver.

Prior to 1855, the McClellan party reported
the presence of placer gold in the Yakima Valley
while searching for a railroad route through the Cascades. As early as 1846, Indians of the Cle Elum
Valley were discovered wearing gold ornaments,
fashioned from placer gold from the headwaters of
the Cle Elum River. Mining of placer gold first took
place in 1855 when it was discovered near Fort Colville on the Columbia River. With the discovery of
gold near Fort Colville, many prospectors rushed to
the area, where as much as $3 to $6 per day could
be made with a gold pan and a rocker. However,
the gold rush was short lived, and prospectors drifted
north into the gold field of British Columbia, or

spread out into the remote areas of northeastern
Washington in search of new gold deposits.

By the early 1870's, lode gold deposits were
discovered in mountains south of Kettle Falls, and
in 1882 mining claims were being staked in the
Calispell Mountain's region of Stevens County. In
1883, the first significant discovery of lead and
silver in Washington was made 2 miles east of Chewelah at Embry camp. Following this discovery,
other discoveries of gold, silver, copper, lead, and
zinc were made, among which was the Eagle that in
1885 was to become the first producing mine of the
area. In 1885, a rich deposit of lead and silver was
discovered and staked as the Old Dominion claim,
6 miles east of Colville on Old Dominion Mountain.
Other discoveries of lead and silver were made in
1885 in the Bossburg district northwest of Colville,
and in the Summit district west of Chewelah. The
discoveries of rich lead-silver ore in the Colville,
Bossburg, Chewelah, and Summit districts of Stevens
County gave birth to silver mining in Washington.
At first, shipments of ore were made to smelters in
Butte, Helena, St. Louis, Denver, and San Francisco.
However, in 1887, Mutual Smelting and Mining
Company of Washington erected a smelter at Colville
and began smelting lead and silver ores from the
neighboring mining districts. Mining camps sprang
up throughout Stevens County, which prior to 1899
included Ferry and Pend Oreille Counties, and many
important discoveries of gold, silver, copper, lead,
and zinc were made.

In the part of Stevens County that later became Pend Oreille County, early-day mining activity
centered around the Metaline Falls area. Placer gold
had been discovered along the Pend Oreille River as

early as 1865, and the first quartz lode claim had been staked in 1873. However, it was not until 1884 that the area was brought into prominence by the discovery of rich lead, zinc, and silver deposits near Metaline Falls on the Pend Oreille River. Among the first mines to be brought into production were the Josephine, which was discovered in 1884, and the Bella May, discovered in 1886. In the southern part of Pend Oreille County, known as the Newport district, deposits of copper, lead, zinc, and silver were found in 1888 near Calispell Lake. However, the discoveries failed to attract attention as they were not as rich as the deposits that had been discovered elsewhere in Stevens County. Prospecting continued in the Newport district; in 1893, the Comstock claim was staked on a zinc-lead-silver-copper deposit near Bead Lake, and the Kootenai Conquest claim was staked on a deposit of lead, silver, and copper near Newport. Later these properties were to become major producers of lead, silver, and copper in the Newport district.

In Ferry and Okanogan Counties, prior to 1886, not much in the way of prospecting and mining had taken place; much of the land had been set aside as Indian reservations that were closed to mineral entry. In 1872, the Colville Indian Reservation was established and consisted of all of Ferry County and the east half of Okanogan County. In 1879, the Chief Moses Reservation was created from that part of Okanogan County that lay west of the Okanogan River. In the spring of 1886, the Chief Moses Reservation was opened to mineral entry and immediately mining claims were staked on deposits of silver, lead, and copper. In the Conconully Lake area, miners organized the Salmon River mining district; 6 miles northeast of the lake the Galena district was organized; and in 1887, miners organized the Ruby district 6 miles south of the lake. Important discov-

eries in the Salmon River district included the Lone Star, Lady of the Lake, Tough Nut, and the Salmon River group. Among the first discoveries in the Ruby district were the Peacock, Ruby, First Thought, Arlington, and Fourth of July. At several mines in the Salmon River and Ruby districts, near-surface ore assayed as high as $1,000 per ton in silver.

Ruby silver camp (1891)

In 1886, silver was also discovered 30 miles north of Conconully in the Nighthawk district; by 1887, quartz veins rich in gold and silver were discovered on Palmer Mountain, as well as east of Palmer Mountain near Wannacut Lake. Among the first mines to produce were the Julia and Kaaba-Texas in the Nighthawk district, and the Black Bear, War Eagle, and Ivanhoe mines in the Palmer Mountain district.

In the Methow Valley, which is 18 miles southwest of Conconully, discoveries of gold, silver, and copper were made in 1887 at the Red Shirt property. By 1892, copper, gold, and minor silver had been discovered in the southern part of the Methow Valley at Squaw and Gold Creeks, as well as near the headwaters of the Twisp River, a major tributary of the Methow river. Prospectors followed the Methow River upstream, and in 1895 staked placer claims on Robinson and Goat Creeks in the Mazama district. However, most lode deposits of copper, gold, and silver were not staked in the district until 1905. By 1895, with the exception of the area north of Winthrop in north-central Okanogan County, almost all of Okanogan County west of the Okanogan River had been prospected.

East of the Okanogan River prospecting remained at a standstill until February 1896, at which time the north half of the Colville Indian Reservation was opened to mineral entry. Following the opening of the reservation, discoveries of gold, silver, and copper were made in the Meyers Creek district east of Oroville. At the present site of Chesaw, a mining camp sprang up, and served as an outfitting point for prospectors and miners. Gold, silver, and copper were discovered in the area surrounding Chesaw, while east of Chesaw on Buckhorn Mountain, deposits of iron, copper, and gold were found. By 1897, prospectors had searched southeast of Chesaw, towards the mining camp of Republic, and discovered deposits of gold and silver in the Wauconda district. The mining camps of Toroda and Bodie sprang up on Toroda Creek, and the gold camp of Wauconda came into existence at the headwaters of Bonaparte Creek.

Until the opening of the north half of the Colville Reservation in February of 1896, no mining claims had been staked in Ferry County; however, prospectors had illegally prospected parts of the county anticipating the opening of the reservation. When the reservation was opened, claims were immediately staked on gold-bearing quartz veins in Eureka Gulch, and by May 1896, the mining camp of Eureka was established. The first claims to be staked were the Iron Mask, Copper Belle, and Lone Pine, followed closely by the staking of the Quilp, Black Tail, San Poil, and Mountain Lion. Not until August 20, 1897, was the Knob Hill gold vein dis-

Eureka gold camp (1897)

covered, which was to become the district's most important deposit of gold and silver. Within a few years, several gold mills were erected in Eureka Gulch, and by 1899 the camp of Eureka, which had been renamed Republic, was booming. Elsewhere in Ferry County, deposits of copper, gold, and silver were discovered in 1897, south of the international border near Danville. Major discoveries of the area included the Lone Star and Washington, 6 miles west of Danville, and the Lucile Dreyfuss, which was 2 miles south of Danville. Also in 1897, copper, gold, lead, and silver were discovered 9 miles northeast of Republic in the Belcher district. Among the first claims were the Belcher and Hawkeye.

In 1898, most of northeastern Washington had been prospected, and important discoveries of gold, silver, copper, and lead had been made. However, the south half of the Colville Indian Reservation, which contained known mineral deposits, was still closed to mineral entry. On June 30, 1898, the south half of the reservation was opened, and prospectors, mainly from the camp of Republic, rushed to the area. Sixteen miles south of Republic, gold, silver, copper, lead, and zinc were discovered at the headwaters of Gold Creek, and the camp of Park City sprang up. Forty miles south of Republic, near the mouth of the San Poil River, deposits of copper, gold, silver, lead, nickel, and cobalt were discovered. On the east bank of the San Poil the settlement of Keller was established. By August 1898, prospectors had moved eastward from Keller to the present site of Inchelium, where deposits of lead and silver were discovered in the Covada district. Although mining claims were staked in 1898 and 1899 in the Covada district, placer gold had been mined in the district as early as 1860 and the Indians of the area had known of the district's metal deposits since 1875.

Keller smelter (circa 1905)

The opening of the south half of the Colville Indian Reservation also opened for mineral entry that part of Okanogan County between the Okanogan River on the west and the Ferry County line on the east. In July 1898, several claims had been staked in the vicinity of the Nespelem Indian Agency, but the discoveries failed to create a rush to the area. Not until 1901, when deposits of high-grade silver ore were discovered at the Apache mine, did the Nespelem district attract many prospectors.

In the spring and summer of 1896, shortly after the opening of the north half of the Colville

Indian Reservation, discoveries of gold were made in the northwest corner of Stevens County near Orient. The best known discovery of the Orient district was the First Thought gold mine, which was to become a leading producer of gold in the county. East of Orient towards the mining camp of Northport, as well as south of Orient toward the mouth of the Kettle River, deposits of lead, zinc, copper, iron, gold, and silver were discovered.

By the turn of the century, valuable metal deposits had been discovered in many parts of Pend Oreille, Stevens, Ferry, and Okanogan Counties. Hundreds of mining claims had been staked in country that was for the most part remote, and accessible mainly by trails made earlier by Indians. Wagon roads had been built between the larger settlements in the region, and over these roads supplies were hauled to the mining camps, and ore was hauled to smelters for refining. In many mining districts, concentrating mills were built to extract gold and sulfide minerals from the ore. Some mining camps grew into towns, whereas other camps were short lived and became ghost camps. Had it not been for the prospectors and miners, many parts of northeastern Washington would probably not have been settled until the late 1800's and early 1900's. Regarding the settlement of mining regions, Trimble (1914, p. 106) states:

> One of the most important permanent improvements, attributable largely to the precious metal product, was the development of agriculture. Prices for all sorts of provisions were very high in the mines and at the towns and stations on the way thither, and this was particularly true with regard to butter, milk, fresh vegetables, etc. . . . . The economic inducement of high prices was needed in order to settle remote valleys, which, but for the miners, would have waited long for settlers.

Elsewhere in eastern Washington, rich deposits of placer gold had been discovered as early as 1860 on Peshastin Creek in Chelan County, and as

Blewett gold camp (circa 1900)

early as 1873 on Swauk Creek in Kittitas County. However, it was not until 1874 that the first lode gold claims were staked in the Blewett and Swauk districts. Although some gold had a high silver content, important discoveries of silver were not made in these districts. In the Lake Chelan area of Chelan County, deposits rich in silver and lead were discovered in the Meadow Creek district in 1891, while beyond the head of the lake and near the headwaters of the Stehekin River, deposits of silver and lead had been discovered in 1885. On July 20, 1896, J. H. Holden discovered a deposit rich in copper, gold, silver, and zinc on Railroad Creek near the head of Lake Chelan. This deposit, would in the future, become one of Washington's major metal-producing mines. Although most mining activity in Kittitas County centered about the Swauk district, by 1881 prospectors had worked their way to the headwaters of the Cle Elum River and discovered deposits of gold, silver, lead, and copper. Deposits of gold, silver, and copper were also found near the headwaters of Gold Creek, and by 1896 small shipments of high-grade ore were being made to the Tacoma smelter. Although several concentrating mills were

built in the Cle Elum and Gold Creek districts, the production of metals from both districts was insignificant.

PRODUCTION

Prior to 1885, the production of silver in Washington came mainly from the refining of placer and lode gold bullion. The bulk of this gold came from the Columbia River placer deposits discovered in 1855, the Similkameen River deposits, discovered in 1859, and the Peshastin-Swauk lode and placer deposits that were discovered around 1873. Mineral production figures for individual counties were not published for years prior to 1903; however, the total gold and silver production in Washington from 1860 to 1884 is reported in the Minerals Yearbook (U.S. Bureau of Mines, 1913, p. 790) as 682,732 ounces of gold and 249,698 ounces of silver.

Excluding silver recovered from the refining of gold bullion, the initial production of silver in Washington began in 1885 when several tons of high-grade lead-silver ore was shipped from the Old Dominion mine to a smelter in San Francisco. This shipment netted $3,000, which provided the incentive needed to develop other silver mines in Stevens County. Shortly after shipments were made from the Old Dominion mine, the production of lead-silver ore began at the Young America, which is 18 miles northwest of Colville and in the Bossburg district. High-grade silver ore stockpiled at mines in Stevens County encouraged Mutual Smelting and Mining Co. of Washington to build a smelter at Colville in 1887. By 1891, ore from mines in the Colville, Kettle Falls, Bossburg, Chewelah, and Summit mining districts were being smelted at the Colville smelter. Elsewhere in Washington, Tacoma Smelting and Refining Co. built a smelter at Tacoma in 1889; in 1890, Puget Sound Reduction Co. built a smelter at Everett, and

North Pacific Reduction Co. constructed one the same year in Spokane.

In the Conconully area of Okanogan County, the first production of silver was in 1889 when high-grade silver ore was shipped to Montana smelters from the Monitor and Fourth of July mines. North of Conconully in the Palmer Mountain district, ore rich in gold and silver was mined at the Black Bear mine in 1887, and at the Black Bear, War Eagle, and Ivanhoe mine in 1890. Ore shipped to Denver from the Ivanhoe netted $636 per ton, mainly in silver. Other major gold-silver producers in 1890 included the Julia, Kaaba-Texas, and Number One mines of the Nighthawk district north of Palmer Mountain. By 1891, gold-silver mines in the Nighthawk and Palmer Mountain districts and lead-silver mines in the Ruby and Conconully districts were making regular shipments to smelters. Mills were built to extract free gold from quartz, while other mills made concentrates from low-grade lead-silver ore that could not be shipped at a profit to distant smelters. In 1892, the price of silver began to fall, and in 1893 a silver panic hit the nation. Production at most mines ceased and within a few months many mining camps were deserted.

Up until 1889, the bulk of silver produced in northeastern Washington came from silver-rich lead ore that was mined mainly in Stevens and Okanogan Counties. However, in 1890, silver recovered from copper ores exceeded that recovered from lead ores, the major producer being the Eagle mine of the Chewelah district in Stevens County. In later years the production of silver from copper ores of the Chewelah district would make the district a leading silver producer of the state.

In 1892, operations at the Colville smelter ceased, and in 1897, after having produced $625,000 in lead-silver ore that contained 40 to 800 ounces in silver per ton, mining operations ceased at the Old Dominion mine. Between 1890 and 1900, an average of 12 mines produced silver in Stevens County from lead-silver and copper-silver ores. The leading mines were the Old Dominion, Young America, Bonanza, Daisy, Silver Crown, Summit, Silver Trail, Deer Trail, Tenderfoot, and Silver Seal. On the average, only 12 mines produced silver yearly, while at least 100 mines were in the assessment or development stages and never reached the production stage.

From 1885 through 1899, silver production in Washington totaled 2,212,376 ounces compared to 231,807 ounces of gold. This represents 9.5 ounces of silver for each ounce of gold, whereas prior to 1885 the ratio had been 2.6 ounces of gold for each ounce of silver. In 1884, only 912 ounces of silver had been produced in Washington, but in 1899, production reached 289,661 ounces. Other metals produced in 1899 included 33,156 ounces of gold, 76,410 pounds of copper, and 1,032,069 pounds of lead, with the bulk of the total state's production coming from Stevens, Ferry, and Okanogan Counties.

By 1900, silver-bearing ore had been discovered in almost all mining districts of northeastern Washington. However, silver constituted the major metal of ore that had been mined in only 22 of over 60 mining districts that had been organized.

EASTERN WASHINGTON, 1900-1970

PEND OREILLE COUNTY

General History

Although mining claims had been staked in the Metaline district as early as 1873, and in the Newport district in 1888, by 1900 there was not yet a major producing mine in the county. The inaccessibility of the region made it one of the most sparsely settled areas of northeastern Washington. In 1905, river-boat transportation into the Metaline district was made possible by the removal of Little Falls, which up until this time has been an obstacle to navigation on the Pend Oreille River. With the removal of Little Falls it became possible to transport ore 40 miles upriver to railheads at Newport, and in 1905, the first shipments of lead-silver ores were made from several mines in the Metaline district. However, it was not until 1910, when Idaho and Washington Northern railroad built a line between Newport and Metaline that mining was firmly established. In 1910, lead mining began at the Josephine mine that had been discovered in 1884; in 1911, the first shipment of zinc ore from Washington was made from the Oriole mine. In 1911, Pend Oreille County was formed from the eastern half of Stevens County.

Grandview mine and mill (circa 1930)

Although zinc mining began in Pend Oreille County in 1911, the county did not become a major producer of zinc until 1929, at which time Pend Oreille Mines & Metals Co. began operations at the Josephine mine. Mining operations at the Grandview began in 1924, and upon completion of a flotation mill in 1929 the Grandview became another major zinc producer in the district. In 1937, American Zinc, Lead & Smelting Co. began lead-zinc mining operations at the Bella May mine that had been discovered in 1886, and by 1937, the major lead-zinc mines of Pend Oreille County were in production. Mining was carried out at the Bella May until 1951, and in 1964, mining ceased at the Grandview mine, which made the Pend Oreille (Josephine) mine the only remaining lead-zinc producer in Pend Oreille County. Since 1964, the Pend Oreille mine has been a constant producer of lead and zinc, as well as minor copper and silver, and currently (1975), is the state's only operating lead-zinc mine.

In the Newport district of Pend Oreille County mining claims had been staked as early as 1888, but not until 1917 was ore shipped from the district. In 1917, 3 tons of lead-silver ore was shipped from the Eagle mine, and other small shipments were made sporadically until 1923 when the mine shut down. In 1923, Bead Lake Gold and Copper Co. produced several hundred tons of copper-lead ore from their Comstock and Conquest claims near Bead Lake. Several hundred tons of ore was produced again in 1928, after which the company suspended operations. The property remained idle until 1960, at which time Lucky Joe Mining Co. reopened the mine and built a small flotation mill; however, after 4 years of minor production, operations ceased. In 1975, no metal mines were operating in the Newport district.

Production

From 1902 through 1969, Pend Oreille County produced $145,944,035 in metallic minerals, the bulk of which was lead and zinc. Of this total, only $14,440 came from mines in the Newport district. Silver produced in the county amounted to 708,330 ounces and was valued at $614,617; almost all silver was produced as a byproduct of lead-zinc mining operations. Although silver has been produced in

Pend Oreille County since 1905, the Metaline and
Newport mining districts are not considered silver
mining districts. Thirty mines in these districts have
produced silver, but only at the Pend Oreille mine
has the total production exceeded 100,000 ounces.
At the Grandview mine, which was the second largest
lead-zinc producer in the county, the total silver
production is reported (Fulkerson and Kingston, 1958,
p. 22) as 76,261 ounces. Only at the Rocky Creek
and Ries (Eagle) mines did the production of silver
exceed that of other metals; however, the total combined production from these mines was only 1,601
ounces.

STEVENS COUNTY

General History

At the turn of the century over 3,000 mining
claims had been staked in Stevens County, but only
12 mines had become significant producers. Along
the Columbia River most rich placer deposits had been
worked out, but some small-scale placer mines were
still operating. The smelter at Colville had shut
down in 1892, but at Northport the Northport Smelting & Refining Co. smelted Canadian copper ore, as
well as small amounts of lead-silver ore from Stevens
County mines. Silver continued to be the county's
most important metal, although the production of
lead, copper, and gold was increasing.

Between 1900 and 1910, the Orient district
became the most productive metal district in the
county. Substantial amounts of gold and silver were
produced at the First Thought mine near Orient,
while copper, gold, and silver came from the Napoleon mine near Boyds. Throughout the rest of the
county other mines produced mainly lead and silver.
In the Bossburg district, the Young America and

Northport smelter (circa 1918)

Bonanza mines were important lead-silver producers,
as was the Old Dominion mine in the Colville district.
On Deep Creek east of Northport high-grade lead
ore was being mined at the Last Chance mine. In
the Chewelah district, United Copper and Copper
King mines were mining and milling copper ore rich
in silver, while at the Eagle and J. Gould mines rich
lead-silver ores were being mined. One of the most
productive silver districts from 1900 to 1910 was the
Deer Trail, where high-grade lead-silver ores were
mined at the Legal Tender, Providence, and Queen
mines. In July 1910, the First Thought mine near
Orient ceased operations, and gold production in
Stevens County declined sharply. However, the
production of silver, copper, and lead remained
about the same as it had been since 1900; the most
productive mining districts were Chewelah, Northport,
and Bossburg.

The year 1912 showed a significant increase
in the production of copper, due mainly to mining
operations at United Copper, the state's leading
copper producer. Other important copper-silver mines
in the Chewelah district were the Amazon, Copper
King, and High Grade. From 1912 to 1920, production of silver in Stevens County averaged around
150,000 ounces yearly.

Between 1916 and 1920, increased mining
operations in the Northport district resulted in an

increase in the production of lead and silver. The leading producers of the Northport district were the Electric Point, and Lead Trust, United Treasure, Great Western, and Byran. Prior to 1916, zinc had not been mined in Stevens County, but in 1916 shipments of zinc ore were made from the Byran mine, and in 1917 zinc was shipped from the Young America mine. These shipments marked the beginning of zinc mining, which later became a major industry in Stevens County.

By 1920, the Northport district had become the county's major zinc-producing district, and in 1923, the Black Rock mine produced 3 million pounds of zinc. Other zinc producers in the Northport district were the Great Western and Northwest mines; however, production from these mines was minor. Although in the early 1920's mining was an important industry in northeastern Washington the mines could not support the smelter at Northport.

In 1922, after 24 years of sporadic operations the smelter ceased operations. Initially the smelter was built to smelt gold, silver, and copper ore from mines near Rossland and Trail in British Columbia; however, starting in 1916, the smelter handled lead ore mainly from mines of the Coeur d'Alene region of Idaho. In addition to ores from Canada and Idaho, minor amounts of ore from several Washington mines were custom smelted at the Northport smelter.

In the mid 1920's, an average of 15 mines operated yearly in Stevens County. Lead continued to be the county's major metal, and the major producers were the Gladstone and Electric Point mines in the Northport district, the Old Dominion mine in the Colville district, and the Cleveland mine in the Deer Trail district. The bulk of the silver produced in 1925 came chiefly from the Old Dominion, which had reopened in 1915, and United Copper, a steady producer since 1906.

United Copper camp (circa 1910)

In 1927, a sharp decline in mining took place in Stevens County when several lead and zinc mines suspended operations. Whereas in 1926, the mines produced $468,646 in gold, silver, copper, lead, and zinc, in 1927 the county's total production of these metals was only $84,340—the lowest it had been since 1903. This decline in production was caused mainly by the depletion of ore bodies at the Old Dominion and Gladstone mines. In 1930, other major producing mines in the county suspended operations, including the Bonanza, which had been in operation since 1907, and United Copper, a leading producer of copper and silver in the Chewelah district since 1906. In the mid 1930's, mineral production increased slightly because of increased production at the First Thought gold mine near Orient, and because of the reopening of the Old Dominion and Daisy mines.

During the depression years of the 1930's, miners began reworking gold placer deposits along the banks of the Columbia River. By 1940, ten placer mining operations were working the river gravels for gold between the Spokane River and the Canadian border; however, the mining operations were small, and yearly production for individual operators seldom exceeded several hundred dollars. Although the 1930's brought about a general decline in lead and zinc production throughout the county, the Electric Point and Gladstone mines were mining about 80 percent of the lead produced in the county. World War II, which started in the late 1930's, brought about an increase in the production of base metals throughout the United States, and by 1940, 14 mines in Stevens County were operating. Among the major producers were the First Thought (gold), Chinto (copper-silver), Electric Point (lead), and Byran (lead).

In 1942, Government Order L-108 haulted gold mining in the United States, which forced the closure of the First Thought mine, as well as other gold mines in Washington. However, the loss in gold production for the county was offset by a large increase in 1941 in the production of zinc at the Blue Ridge mine in the Northport district. Further increases in zinc production occurred in 1944, when the Deep Creek mine in the Northport district began mining operations. In addition to zinc, these mines produced considerable amounts of lead and silver. Other major producing mines were the Bonanza and Old Dominion, both of which were lead-silver mines. At the Old Dominion, mining was confined to old dump material, and the reworking of tailings from early-day mining operations.

The end of World War II in 1945 created an industrial demand for lead and zinc, which resulted in an increase in mine production for Stevens County. By 1947, metal production in the county had risen to $1,251,190, which, except for 1917, was the greatest it had been since 1902. The Northport district

Admiral mine and mill (circa 1950)

was the most active district and produced 80 percent of ore mined in Stevens County. Major producing mines were the Deep Creek (lead-zinc), Last Chance (lead-zinc), Admiral (zinc), Gladstone (lead), and Electric Point (lead). In the Bossburg district the Bonanza mine operated a 100-ton-per-day flotation mill that produced silver, copper, lead, and zinc concentrates, while in the Deer Trail district a 75-ton flotation mill at the Cleveland produced silver, lead, and zinc concentrates. However, in 1948, mining at the Cleveland mine, which had begun in the late 1890's, ceased. After 1947, metallic mineral production in Stevens County increased steadily, and in 1955 production reached $6,181,852. Zinc accounted for 70 percent of the production, followed in order of decreasing value by lead, copper, silver, and gold. Major producing mines at this time were the Van Stone zinc mine, which had begun mining operations in the Northport district in 1952, Lead Trust, Deep Creek, and Admiral mines. Elsewhere in the Northport district, minor amounts of lead and zinc were produced at the Gladstone, Red Top, Last Chance, Electric Point, and Morning mines.

In other districts of the county, mining ceased and many mines stood idle or were abandoned; by 1958, mining operations at all major lead-zinc mines had ceased. The production of gold, silver, copper, lead, and zinc in Stevens County dropped from a high of $1.7 million in 1957 to a low of $11,000 in 1958. Small shipments of silver-lead-zinc ore were made from several mines in the years that followed, and exploration work was undertaken at the Bonanza mine by the Bunker Hill Co., and at the Anderson mine by American Zinc. In 1964, American Smelting and Refining Co. reopened the Van Stone mine, which had closed in 1955; in 1966, American Zinc Co., after building a 1,400-ton-per-day flotation mill, placed the Calhoun mine into production. The combined production from these two mines was several

million dollars yearly. In addition to zinc, the mines produced minor lead, copper, silver, and gold. Low zinc prices in 1968 resulted in the closure of the Calhoun mine, leaving the Van Stone the only major operating lead-zinc mine in Stevens County. In 1969, attempts were made by Arco Industries to place the Daisy-Tempest mine, near Daisy, and the Bluebird mine, near Kettle Falls, into production. For several months, copper, silver, and lead ores from these mines were concentrated at a 200-ton flotation mill near Colville; however, mining was not profitable, and operations ceased. In 1970 and 1971, small amounts of silver-lead-zinc ore were mined by Silver Crown Mining Co. at the Casteel and Hubbard mines near Northport. After milling the ore at a 100-ton flotation mill, concentrates were shipped to the Trail smelter; however, after making several small shipments, mining and milling operations ceased in late 1971. Early in 1971, the Van Stone mine shut down when ore in their open pit was exhausted. Thus, mining, which had begun in Stevens County in 1885, had for all practical purposes, at least temporarily, come to an end.

PRODUCTION

According to the U.S. Bureau of Mines, from 1902 to 1971 no fewer than 200 metal mines at one time or other operated in Stevens County. Production of gold, silver, copper, lead, and zinc totaled around $59.4 million, the distribution of which is as follows:

Zinc	$39,004,830
Lead	14,002,953
Copper	2,638,005
Silver	2,414,122
Gold	1,369,801

Although as many as 200 metal mines were in operation from 1902 to 1971, at only 19 mines did the

TABLE 6.—Major mines of Stevens County

Property	Total tonnage	District	Chief metals
Admiral	22,464	Northport	Zinc, lead
Black Rock	17,771	Northport	Zinc, lead
Blue Ridge	56,217	Northport	Zinc, lead, copper, silver
Bonanza	101,994	Bossburg	Lead, zinc, silver
Calhoun	900,000	Northport	Zinc
Cleveland	26,459	Deer Trail	Lead, silver
Copper King	13,027	Chewelah	Copper, silver
Deep Creek	763,307	Northport	Lead, zinc
Electric Point	51,200	Northport	Lead
First Thought	81,621	Orient	Gold, silver
Gladstone	19,942	Northport	Lead
Last Chance	11,529	Northport	Lead, zinc, silver
Lead Trust	6,323	Northport	Lead, zinc
Loon Lake	7,317	Springdale	Copper
Napoleon	157,728	Orient	Copper, gold, silver
Old Dominion	4,088	Colville	Silver, lead
United Copper	355,132	Chewelah	Copper, silver
Van Stone	8,200,000	Northport	Zinc, lead
Young America	13,389	Bossburg	Lead, zinc, silver

total production exceed 4,000 tons. These properties, along with total tonnage of ore mined, chief metals, and districts are shown in table 6.

Production figures published by Fulkerson and Kingston (1958, p. 14–19) for 1902 through 1956 show that mines of the Northport district produced 75 percent of metals in terms of dollars mined during this interval. During this same interval, mines of the Chewelah district produced 61 percent of the silver. A breakdown for the dollar value of metals produced by the districts, as well as ounces of silver produced is shown in table 7.

TABLE 7.—Metal production in Stevens County, 1902-1956

Mining district	Total metal production[1]	Total silver (troy ozs)
Northport	$29,150,762	182,916
Bossburg	3,926,068	307,288
Chewelah	3,642,116	1,746,467
Orient	1,158,297	32,020
Deer Trail	396,372	188,465
Colville	373,472	362,467
Kettle Falls	87,741	54,115

[1] Includes gold, silver, copper, lead, and zinc.

FERRY COUNTY

General History

At the turn of the century, mining activity in Ferry County centered about the Republic district where mining was carried out at the Quilp, Lone Pine-Surprise, Black Tail, San Poil, El Caliph, Ben Hur, and Republic mines. However, with the closure of a large custom mill in the district in 1901, mining slumped.

Republic mill (circa 1900)

In 1902 and 1903 two railroad companies extended their lines to Republic, which made it possible to ship crude gold ore to distant smelters. However, because the ore proved to be highly siliceous, and because of high railroad rates, mining once again declined. In 1909, freight rates were reduced, and the mines resumed operations. Production for 1909 was 10,179 ounces of gold and 53,677 ounces of silver, most of which came from the Republic, Quilp, Mountain Lion, and Lone Pine mines. In 1910, mining operations began at the Knob Hill mine, and 9 mines in the district produced 37,874 tons of ore that averaged $25.71 per ton in gold and silver. The Republic mine was the district's leading producer,

having produced $1,400,000 in gold and silver since 1897.

The year 1911 was a peak production year for Ferry County. Thirteen mines, eight of which were in the Republic district, produced $896,848 in gold, silver, and copper. Ore from the Republic mines averaged $22.52 in gold and silver, and contained about 4 ounces of silver for each ounce of gold. Full-scale production of copper at the Lone Star and Washington began in 1911, and continued through 1918. By 1912, total production for the county reached $5 million, of which 83 percent was gold. Small shipments of copper ore had been made from the Lone Star and Washington mine in the Danville district, and several small shipments of silver-lead ore had been made from mines in the Covada district.

Copper mining was undertaken in 1916 at the Talisman mine in the Orient district, as well as at the Belcher mine in the Belcher district from 1913 to 1917. Since 1903, several mines in the Belcher district had been shipping gold-silver-copper-bearing ore to smelters in Canada. The total copper production in the county, from 1911 to 1918, amounted to $428,351, which represents 88 percent of the county's copper production from 1903 to 1950.

By 1920, mining had declined considerably, and only seven mines in Ferry County were operating. In the Republic district the Quilp, Last Chance, San Poil, and Knob Hill mines produced only 12,343 tons of ore that averaged $11.53 a ton in gold and silver. In the Covada district, small shipments of silver ore were made from the Longstreet and several other mines. In the Keller district, development work was underway at the Shamrock mine, which had produced silver, lead, and zinc in 1914, as well as at the Manila mine where copper ore was being stockpiled. Although a 100-ton copper smelter had been built at

Last Chance mine (circa 1900)

Knob Hill mine (1950)

Keller by Keller and Indiana Consolidated Smelting Co., it was never placed into operation.

By 1925, mining in the county was stagnant. In the Republic district, only the Knob Hill and Quilp mines were operating, and only $162,541 in gold and silver was produced. No ore was milled in the district, but was shipped as crude ore to the Trail smelter because of its high silica content. By 1930, mine production in the county had dropped to an all-time low of $10,177. Because there was no longer a demand for siliceous ores at smelters, most large gold producers in the Republic area had been idle since 1928. In 1934, when gold rose from $20.67 to $35 per ounce, gold production increased to $161,491 and silver to $15,589, making Ferry County the state's leading silver producer. In addition to 21 producing lode mines, which produced gold, silver, copper, and lead, 15 placer mines operating along the shores of the Columbia River produced 104 ounces of placer gold. In 1937, gold and silver production in the Republic district increased sharply when Knob Hill Mines, Inc. placed into operation a 400-ton-per-day cyanidation plant. Further improvements in the recovery of gold were made in 1940, when flotation units were added to the plant. In 1940, the district produced $849,448 in gold and silver from siliceous

ores; 14 lode mines were operating, 12 of which were in the Republic district. Mines in the Republic district produced 179,983 tons of ore, compared to 201 tons from the Danville district and 29 tons from the Covada district.

From 1940 to 1950, an average of only five mines were active in the county, with the bulk of gold and silver coming from the Knob Hill mine. The average yearly production was around 20,000 ounces of gold and 100,000 ounces of silver. Outside of the Republic district, some copper, lead, and zinc was mined from 1948 through 1950 at the Talisman mine in the Orient district. Since 1950, the Knob Hill mine has been the only major producing mine in Ferry County, as well as the state's major gold producer.

Production

The estimated production of gold, silver, copper, lead, and zinc for Ferry County from 1896 to 1970 is around $48.3 million. Gold accounts for 85 percent of the total, silver 13 percent, copper 1.9 percent, and lead and zinc less than 1 percent. The bulk of gold and silver produced in Ferry County came from the siliceous gold ore of the Republic

TABLE 8.—Major mines of Ferry County,
1900–1970

GOLD AND SILVER	
Republic district	Republic Princess Maude Quilp Surprise Lone Pine Pearl Last Chance Ben Hur Knob Hill Tom Thumb Mountain Lion Morning Glory El Caliph California San Poil Trade Dollar
Belcher district	Belcher Copper Key
Danville district	Lucile Dreyfus
SILVER	
Covada district	Gwin Longstreet Meteor
Sheridan district	Sheridan Zalla M.
COPPER	
Danville district	Lone Star and Washington
Orient district	Talisman

district. Copper came mainly from the Danville district, and lead and zinc from the Covada and Keller district. Between 1904 and 1957, Ferry County produced 5,917,851 ounces of silver. Prior to 1904, production figures for individual counties were not published. However, the U.S. Bureau of Mines estimates silver production of the Republic district from 1896 to 1903 at 558,288 ounces. A rough estimate of silver produced from 1957 to 1970, would be around 3.5 million ounces, making the county's total silver production almost 10 million ounces. Mines that were major producers in Ferry County, between 1900 and 1970, are shown in table 8.

OKANOGAN COUNTY

General History

By 1900, most of Okanogan County had been prospected, and several thousand mining claims staked. About a dozen mines were making small shipments of gold, silver, and lead ore to smelters, while around 150 properties were in development stages. In the Conconully district, the Arlington, Fourth of July, and Nevada mines were mining silver ore; however, many silver mines in the district had remained idle since the silver panic of 1893. In the Palmer Mountain and Wannacut Lake districts, several concentrating mills had been built, and gold-silver concentrates were being shipped to the Tacoma smelter from Golden Zone, Black Bear, War Eagle, Ivanhoe, Triune, and Horn Silver mines. In the Methow district, the Hidden Treasure and Hunter mines on Squaw Creek were producing ore that contained gold, silver, and copper, while in the Wauconda district development work was underway at the Wauconda and Bodie prospects. In the Sheridan district, which was east of the Wauconda district, small shipments of high-grade silver ore were being made by the Zalla M., Sheridan, and American Flag mines. In the Meyers Creek district east of Oroville, mining in 1900 was at a standstill; four small amalgamation mills had been built in the district, but the mills proved to be inefficient due to the pyritic nature of the ore. Prior to 1900, the district had produced around $100,000 in gold, $40,000 of which came from placer deposits on Mary Ann Creek.

From 1900 to 1923, an average of only eight mines operated yearly in Okanogan County; the most productive districts being Conconully, Meyers Creek, and Palmer Mountain. Silver was the most valuable metal produced, followed in order of decreasing value by gold, copper, and lead. The only significant producer of silver in 1925 was the Apache mine near Nespelem. Several shipments of high-grade silver ore had been made since 1911; some ore contained as much as 300 ounces of silver per ton. Between 1925 and 1934, several small gold mines operated, but the yearly combined production from the mines never exceeded $7,000.

Because of the increase in the price of gold in 1934, 12 of the county's gold mines resumed production. Major producing mines were the Bodie, Poland China, Hiawatha, Josie, Spokane, and Ruby.

Ruby mine (1920)

Whereas in 1933, only $13,785 in gold was produced in Okanogan County, in 1936, gold production rose to $125,979. The bulk of the gold came from the Bodie mine, where nearly 19,000 tons of ore was mined and milled in a 70-ton amalgamation and gravity-concentration plant. Mining operations were suspended at the Bodie mine early in 1938, which resulted in a sharp decline in the county's gold production; however, this decrease in gold production was offset by an increase in the production of silver, the bulk of which came from the Arlington mine near Conconully. In 1938, the county's mines produced 85,731 ounces of silver, as compared to 1,120 ounces of gold.

In 1939, production of gold in Okanogan County reached $270,130, which was to be an all-time high for the county. The increase in gold production was due mainly to increased production of the Alder mine near Twisp, where mining had begun in 1937. In addition to the Alder, 26 other mines in Okanogan County produced copper, gold, and silver in 1939. In mid-1942, mining operations at the Alder ceased, which resulted in a drastic drop in the county's gold production. Whereas, in 1942 gold production totaled $10,500, in 1943 it totaled only $70. However, 1943 marked the beginning of significant lead-zinc production in Okanogan County. Whereas prior to 1943, not much more than $3,000 in lead had been produced in any given year, $22,800 in lead, as well as $11,448 in zinc was produced in 1943. The bulk of the production came from the Kaaba-Texas

Kaaba-Texas mine and mill (1938)

mine in the Nighthawk district. From 1943 to 1951, Kaaba-Texas mine was the major silver-copper-lead-zinc producer in the county. Gold mining resumed at the Alder mine in 1950, and until 1953 the mine was once again the county's leading gold producer. In addition to gold, the mine produced substantial amounts of silver, copper, lead, and zinc. When the mine closed in 1953, it had produced around $850,000.

In the Conconully district most lead-silver mines had shut down when the rich near-surface ores had been mined out. Mining in the district resumed again in 1954 at the Mohawk and Peacock mines, but operations were suspended in 1957 due to the low lead and silver content of the ore.

The last attempt at silver mining in the Conconully district took place from 1960 to 1964, at which time small shipments of high-grade silver ore were made from the Fourth of July mine on Ruby Hill to the Trail smelter in Canada. In 1965 and 1968, mining on a small scale took place at the Andy O'Neil mine near Nespelem, and small amounts of lead-silver ore was shipped to the Kellogg smelter in Idaho. The closure of the Andy "O" in 1968 brought an end to mining, for the time being, in Okanogan County. Although there have not been any producing mines in Okanogan County since 1968, exploration of the county's metallic mineral deposits takes place almost every year. In recent years, extensive exploratory work has taken place in the Mazama, Lost Lakes, Palmer Mountain, Meyers Creek, and Conconully districts.

Production

Production of gold, silver, copper, lead, and zinc in Okanogan County from 1889 to 1968 is estimated at $3,217,525. For the most part, the bulk of the silver came from lead-silver and gold-silver ores, whereas only minor silver came from copper and zinc ores. Total silver produced in the county is about 1,000,000 ounces. Of this total 954,671 ounces was reported by the U.S. Bureau of Mines as being produced between 1903 and 1968. The remaining 146,329 ounces is the estimated production between 1889 and 1903. Total silver production, compared to gold, copper, lead, and zinc production follows:

Gold	$1,505,359	53,236 ounces
Silver....	885,329	1,000,000 ounces
Copper...	394,031	2,103,332 pounds
Lead	287,576	2,717,185 pounds
Zinc	145,230	1,444,460 pounds

Although over 100 properties in the county's 16 mining districts have a record of production, only 34 mines are considered major producing mines. The bulk of the gold came from the Palmer Mountain and Methow-Squaw Creek districts; most silver came from the Conconully and Nespelem district; most copper came from the Palmer Mountain and Twisp districts; and almost all the lead and zinc produced in the county was from the Nighthawk district. Major producing districts and the major mines are shown in table 9.

Silver miners (circa 1900)

TABLE 9.—Major producing mines of Okanogan County, 1900-1970

GOLD	
Palmer Mountain district	Black Bear Bullfrog Pinnacle
Methow- Squaw Creek district	Hidden Treasure Highland Holden-Campbell Methow (London) Friday
Twisp district	Alder Red Shirt
Meyers Creek district	Poland China Reco Gray Eagle
Wannacut Lake district	Triune
Oroville district	Okanogan Free Gold
Wauconda district	Bodie

LEAD AND ZINC	
Nighthawk district	Four Metals Kaaba-Texas
Park City district	Mountain Boy

SILVER	
Conconully district	Arlington First Thought Last Chance Fourth of July Tough Nut Leuena
Nespelem district	Apache Little Chief
Palmer Mountain district	Ivanhoe
Nighthawk district	Ruby Horn Silver
Sheridan district	Sheridan

COPPER	
Palmer Mountain district	Copper World Extension
Oroville district	Golden Chariot
Conconully	Blue Lake
Twisp district	Alder

CHELAN COUNTY

General History

At the turn of the century, mining in Chelan County centered about the Blewett mining district. Several small stamp mills had been erected in Culver Gulch, the largest of which was the 20-stamp mill of the Blewett Gold Mining Co. Among the main producing mines in the gulch were the Tip Top, Pole Pick, Peshastin, Culver, and Blewett. Ore from the mines averaged only $3 to $10 per ton in gold; however, ore containing up to several thousand dollars per ton in gold was occasionally mined. Whereas in the Republic district the gold-silver ratio was 1:4, the gold-silver ratio of Blewett ore was 4:1. By

Blewett mill (1904)

1903, only three gold mines in the district were operating, while 51 properties were in assessment or development stages.

In 1903, gold was discovered in Crum Canyon of the Entiat district. Ore from the Rex and Ethel mines averaged $40 per ton in gold, while some ore ran as high as $1,700 per ton. In the Stehekin district very little work had been done on lead-silver deposits, which were discovered in 1885 in the Horseshoe Basin area at the headwaters of the Stehekin River. From the head of Lake Chelan, this remote area near the crest of the Cascades was accessible only by 30 miles of poor trail. In 1905, Horseshoe Basin Mining and Development Co. and Cascade Gold and Copper Co. purchased most claims in the basin and began development work.

From 1900 through 1916, production from gold mines in Chelan County averaged around $10,000 yearly, almost all of which came from mines in the Blewett district. From 1917 through 1928, mining in the county was stagnant with not much more than 50 tons of ore being produced yearly. In 1929, Howe Sound Mining Co. acquired the Holden property on Railroad Creek, in the Chelan district, and began an extensive development program. Also in 1929, Royal Development Co. erected a 200-ton flotation mill at their copper prospect on Phelps Ridge in the Chiwawa district. Low-grade copper ore was mined and milled in 1930, 1935, 1936, and 1937; however, operations ceased in 1937 when mining proved unprofitable.

In 1938, full-scale mining began at the Holden mine, and Chelan County became the state's major producer of gold, silver, and copper. In the Wenatchee area, American Smelting and Refining Co. mined ore at the Golden King mine in 1938 and

Holden mine (1937)

1939, and in 1944, Knob Hill Mines, Inc. took over the property and produced gold ore until 1946. However, not until the Lovitt Mining Co. took over the property in 1949 did the Golden King become a successful gold mine. In 1950, most gold, silver, and copper produced in Washington came from the Holden and Golden King mines. The closure of the Holden in 1961 left Washington without a major copper-producing mine. In 1967, mining operations at the Golden King mine ceased, which left the state with only one major gold producer—the Knob Hill mine at Republic. Thus, in 1967, the production of precious and base metals in Chelan County, which had begun in 1874, came to an end.

Production

Chelan County has been one of the leading producers of metals in Washington. According to production figures published by the U.S. Bureau of Mines, the county produced $73.75 million in gold, silver, copper, lead, and zinc, from 1903 to 1965.

Gold	$33,460,800
Silver	2,190,093
Copper	32,996,349
Lead	5,894
Zinc	5,103,344
Total	$73,756,480

Production figures for years prior to 1903 have never been published; however, Weaver (1911, p. 71) estimates the gold production from 1874 to 1901 at $1,500,000, almost all of which came from the Blewett district. From 1870 to 1935, production of silver in the county was minor. Yearly production seldom exceeded 100 ounces, and in 1912, it reached a high of 823 ounces, all of which came from gold ore. However, in 1935, with the opening of the

Royal copper mine in the Chiwawa district, silver production rose to 2,884 ounces; in 1936, the mine produced 7,694 ounces of silver, and in 1937, it produced 4,282 ounces. When the Holden mine went into production in 1938, the production of gold, silver, and copper in Chelan County reached an all-time high. The bulk of the production in the county came from the Holden and Golden King mines, which in 1938, produced a total of 31,525 ounces of gold, 124,590 ounces of silver, and 11,861,000 pounds of copper.

From 1938, until 1958, the Holden mine produced a total of 212,000,000 pounds of copper, 40,000,000 pounds of zinc, 2,000,000 ounces of silver, and 600,000 ounces of gold. The combined production of $66.5 million made it one of the richest metal deposits in the state. Total production through 1964 from the Golden King mine was around $13.5 million, and consisted of about 367,500 ounces of gold and 420,000 ounces of silver. Although there have been many small producing mines in Chelan County, very few mines produced in excess of $50,000.

TABLE 10.—Major mines of Chelan County, 1874-1964

GOLD	
Wenatchee district	Golden King
Entiat district	Rex Rogers
Blewett district	Culver Black Jack Peshastin North Star Pole Pick Blewett Tip Top
Chelan district	Holden
Chiwawa district	Royal (Red Mountain)

KITTITAS COUNTY

General History

In Kittitas County, around 1900, mining activity centered about the Swauk mining district where placer gold deposits had been discovered in 1873. Many miners had sold their claims to large companies, hydraulic mining had replaced sluice boxes and rockers, and Liberty and Meaghersville had become the district's main mining camps. Although most gold came from placer-mining operations, several lode deposits were being mined on a small scale. No more than a dozen mines produced placer gold yearly, and seldom did the production exceed $5,000

Arrastra at Liberty (1924)

dollars. By 1916, Liberty had become a deserted town, for most miners had moved to Meaghersville where the settlement of Liberty presently stands.

In 1922, Swauk Mining & Dredging Co. placed a gold dredge into operation at the confluence of Swauk and Williams Creeks; however, operations were soon suspended because mining was not profitable—the dredge was too small to handle large boulders. In 1926, Kittitas Gold Mining Co. placed into operation the largest gold dredge that ever operated in Washington. Because of numerous problems, the operation lasted only around $2\frac{1}{2}$ months.

With a rise in the price of gold from $20 to $35 in 1934, attempts were once again made to place several lode and placer gold mines into operation. A slight increase in gold production occurred in the district, but within several years most mines were again inactive. In 1940, Clear Creek Dredging Co. of California placed a large portable washing plant into operation on Swauk Creek, and produced over $50,000 in gold. Like other placer mining operations this operation also failed, and in 1941, only $665 in gold was produced in the district.

From 1941 until 1960, except for several small-scale placer mining operations, gold mining in the Swauk district was at a standstill. Although most lode gold mines in the district yielded very little gold, the gold discovered by Clarence and Ollie Jordin on Flag Mountain is worthy of mention. Beginning in 1932, the brothers had discovered several pockets of crystalline gold, containing up to several hundred ounces of gold. In July 1950, at the Ace of Diamonds mine, Clarence Jordin discovered his largest pocket, which contained around $10,000 in crystalline gold.

The last attempt to mine placer gold on a large scale in the Swauk district began in 1960, when Golden Thunderbird Mining Co. erected a portable washing plant on Williams Creek near Liberty. Many

TABLE 11.—Silver production of eastern Washington's major silver-producing counties, 1903-1956

Year	County					
	Pend Oreille	Stevens	Ferry	Okanogan	Chelan	Kittitas
	Troy ounces					
1903	...	101,567	157,478	38,973	...	13
04	...	70,226	38,745	10,792	...	...
05	...	26,196	34,945	10,690	613	c
06	45	6,327	26,463	6,558	2	c
07	...	17,100	14,321	3,612	53	106
08	...	53,296	7,906	25,526	96	109
09	...	21,569	53,694	3,248	117	89
1910	...	13,281	185,357	3,317	137	57
11	1,602	40,664	185,239	12,204	480	55
12	...	233,535	152,336	23,999	823	36
13	...	148,167	162,596	16,420	189	61
14	...	91,480	163,674	8,039	28	71
15	...	109,750	103,472	41,852	84	67
16	85	174,612	138,293	18,796	177	76
17	1,552	161,013	98,080	14,617	10	97
18	394	168,669	101,376	32,110	151	38
19	288	156,562	65,684	28,893	2	15
1920	...	134,037	33,274	17,007	...	15
21	...	75,189	29,099	38,017	6	46
22	220	103,108	32,725	68,312	4	68
23	509	87,560	97,991	30,926	...	19
24	1,291	132,931	70,347	477	...	11
25	1,043	108,206	41,710	1,479	1	5
26	1,818	90,693	59,160	188	...	452
27	2,913	19,952	114,863	...	5	5
28	721	22,398	65,198	185	...	30
29	2,250	22,598	3,628	4	316	589

Continued on next page

Note: Prior to 1903, silver production was not reported on an individual county basis. After 1956, silver production not reported to avoid disclosing confidential company information.

... denotes no production; c denotes production but concealed with other counties.

TABLE 11.—<u>Silver production of eastern Washington's major silver-producing counties, 1903-1956</u>—Continued

Year	County					
	Pend Oreille	Stevens	Ferry	Okanogan	Chelan	Kittitas
	Troy ounces					
1930	1,219	19,454	1,542	26	1,107	18
31	4,595	9,303	7,021	68	14	24
32	2,606	656	13,521	71	36	142
33	3,263	1,043	13,020	894	26	151
34	1,151	17,724	24,114	568	25	76
35	...	17,920	26,983	3,506	2,894	78
36	3,317	7,982	36,324	9,854	7,694	62
37	12,587	14,221	82,238	9,607	5,161	283
38	14,584	31,561	119,688	85,731	124,590	17
39	11,603	35,786	159,670	44,522	184,059	152
1940	8,609	22,372	110,579	13,949	199,343	520
41	10,807	22,147	173,392	1,741	175,891	3
42	9,630	17,460	164,987	457	174,790	...
43	7,335	15,830	173,115	31,095	141,930	...
44	15,580	16,830	132,075	53,100	103,694	...
45	11,191	6,667	117,599	24,314	121,434	...
46	7,375	21,516	103,260	27,964	93,968	5
47	10,674	58,253	133,053	42,673	48,968	...
48	9,521	44,992	179,295	4,645	137,242	...
49	11,396	48,604	153,429	7,635	135,662	...
1950	20,432	47,973	152,671	5,055	137,483	32
51	22,896	44,166	c	15,007	113,155	...
52	29,910	40,319	c	3,189	c	28
53	34,574	29,702	c	4,993	c	c
54	22,616	15,682	c	c	c	...
55	26,329	43,865	c	c	c	20
56	29,373	22,303	c	11,501	78,355	309
Total	357,904	3,065,017	4,315,230	858,406	1,990,815	4,050

Swauk Creek gold dredge (1926)

nuggets were recovered, the largest of which was 6 ounces; however, not enough gold was recovered to make the venture profitable. The company was reorganized in 1965 as Gold Placers Inc., and equipment capable of moving larger volumes of gold-bearing gravels was used. This placer mining operation like others in the past also proved unprofitable. Mining elsewhere in Kittitas County from 1900 through 1973 was insignificant. Small shipments of copper, gold, and silver ores were made from several mines at the headwaters of the Cle Elum River, but the combined production of the mines did not exceed $10,000.

Production

The estimated production of gold, silver, copper, and lead in Kittitas County from 1884 to 1970 is $2,251,729, and is distributed as follows:

Gold	Silver	Copper	Lead
$2,227,169	$22,863	$1,639	$58

The bulk of the gold came from placer mines that operated in the Swauk district prior to 1903. From 1884 through 1902, around $2 million in gold and $20 thousand in silver was produced in Kittitas County. Most silver came from the refining of placer gold, much of which contained 25 percent silver.

WESTERN WASHINGTON, 1874-1970

General History

On the western slopes of the Cascade Mountains, placer gold deposits had been worked as early as 1868 in the Sultan Basin area of Snohomish County, and as early as 1875 on Slate Creek in Whatcom County. However, it was not until the late 1880's and early 1890's that most of the silver-bearing lode deposits of western Washington were discovered. The first discoveries of silver were made in 1874 in the Silver Creek district of Snohomish County. A small rush followed, but it was not until 1882, at which time a trail was built from Index to the townsite of Galena at the mouth of Silver Creek, that prospecting in the district reached its peak. Many mining claims were staked on deposits of silver, lead, and gold that were discovered along Silver Creek, and in 1890, ore that assayed 30 percent lead and 350 ounces per ton in silver was shipped to a Denver smelter. This initial shipment was followed by other small shipments in 1892, consisting of ore rich in lead, silver, and gold from the Vandalia, Idaho, and Billy Lee mines. Although numerous metal deposits were discovered in the Silver Creek district the deposits proved to be small, and no major mines developed.

One mile north of the headwaters of Silver Creek, and in a basin at the headwaters of the south fork of the Sauk River, Joseph Pearsall discovered a ledge of galena on July 4, 1889, which he named the "Independence of 1776." This proved to be the initial discovery of the Monte Cristo mining district. Prospectors rushed to the district, and several hundred claims were staked; at the confluence of "76" Creek and Glacier Creek, the mining camp of Monte Cristo sprang up.

WORKS OF THE

Tacoma Smelting & Refining Co

Tacoma, Wash.

Tacoma smelter (1890)

Although many discoveries were dug upon by the prospectors, extensive development of the district's mineral deposits did not take place until 1891, at which time the Rockefeller syndicate purchased controlling interest in the Monte Cristo, Pride of the Mountains, and Rainy claims, and organized the Monte Cristo Mining Co. In 1890, the syndicate built a smelter at Everett, and in 1892 and 1893 they constructed a railroad from Everett to the mines at Monte Cristo. A 300-ton-per-day concentrating mill was built at Monte Cristo to handle ore from the Mystery Hill and Pride of the Mountains mines. These mines became the major producing mines of the district, as well as the first major producing metal mines in western Washington. Around 250 men were employed at the two mines and concentrating mill, and an average of 1,200 tons of gold-silver concentrates was produced monthly. From 1890 through 1908, the district produced around 334,000 tons of ore that

averaged 0.40 ounce of gold and 5 ounces of silver per ton. The greatest share of this production came from the Mystery Hill and Pride of the Mountains mines. Other producing mines were the Justice, Foggy, O & B, Peabody, and Rainy. In January 1907, railroad service to Monte Cristo was discontinued and by 1908, the concentrator, as well as most mines, had shut down; this brought mining in the Monte Cristo district to an end. Inasmuch as there was no longer a need for the smelter at Everett, it was sold to American Smelting and Refining Company, and dismantled. Metals produced in the Monte Cristo district from 1890 through 1907 were valued at around $3.67 million. Although the dollar value of gold exceeded that of silver, the district produced 1,670,000 ounces of silver compared to 133,600 ounces of gold. The main factor that forced the mines to close was the declining gold and silver content of the ore. When the mines first began production in

Everett smelter (1900)

1890, the ore averaged $25.20 per ton in gold and silver. As mining progressed, the average grade of the ore fell to $16.20 per ton, and in 1907 it had dropped to only $11.80 per ton. Inasmuch as mining, concentrating, freight, and smelting charges totaled $18 per ton, the mines were no longer profitable to operate, and Monte Cristo became a deserted mining camp. Elsewhere in Snohomish County most of the early discoveries did not prove to be valuable, and the claims were abandoned by the prospectors who discovered them.

In 1891, two years after metals had been discovered in the Monte Cristo district, deposits of gold, silver, copper, and lead were discovered in the Silverton area, which was 9 miles northwest of Monte Cristo. Among the first claims to be staked were the Hoodoo, Independent, Anacortes, and Bonanza

Queen. The mining camp of Independence was established, but in August 1891, the camp changed its name to Silverton. The Stillaguamish mining district was organized, but by 1896, it had become the Silverton district. Hundreds of claims were staked with the hopes that the district would be as rich as the Monte Cristo district; however, by the early 1900's most prospectors and miners had left the district. None of the discoveries proved to be large, and by 1918, only around 1,000 tons of ore had been shipped from the Silverton district. The largest producer in the district was the Bonanza Queen, which by 1918 shipped 830 tons of ore containing 3 percent copper and 3 ounces per ton in silver.

By 1891, prospectors had worked their way into the Sultan Basin area and discovered high-grade silver deposits at the headwaters of Williamson Creek.

Sunset mine (circa 1935)

The Magus, Dee Pree, and Hard to Beat claims proved to be the most valuable, and the claim owners merged in 1897 to form the Forty-Five Consolidated Mining Co. From 1896 through 1901, ore valued at $100,000 was shipped to the Everett smelter from the claims. The average silver content of the ore was 90 ounces per ton, while some shipments averaged 170 ounces per ton in silver. After 1901, mining did not prove profitable because most of the ore contained less than $15 per ton in gold and silver.

Discoveries that had been made in the Index area of Snohomish County in 1872 were explored and found to contain copper that was accompanied by minor gold and silver. Among the early producing mines were the Sunset, Ethel, Copper Bell, and

Wilbur Index. The Sunset became the only major mine in the Index district; and from 1902 through 1940, the mine produced 7,000 tons of copper, 156,000 ounces of silver, and 1,500 ounces of gold. Although significant amounts of silver were produced at the Sunset, the ore averaged only 0.59 ounce of silver per ton.

In King County, deposits of gold, silver, copper, iron, and lead were discovered in the Snoqualmie district in 1891, and in the Miller River district in 1892. Early prospecting activity in the Index and Miller River districts was influenced largely by the construction of the main transcontinental line of the Great Northern Railway eastward up the Skykomish Valley in 1892. The railroad made the area more accessible, and prospecting became active over the entire region. A small railroad station was built near the present site of Index, and the station became the supply camp for prospectors and miners. Early producing mines in the Miller River district were the Apex, Seattle-Cascade, Cleopatra, and Coney Basin. The Apex proved to be the richest, and by 1900 had produced 300 tons of gold ore worth $80,000; by 1943, the production of the mine reached $300,000. Al-

though several of the more promising deposits of the Snoqualmie district were explored and developed, production from the district was insignificant. Prior to 1900, several small shipments of copper ore were made from the Dutch Miller mine; and between 1952 and 1956, copper ore was shipped to the Tacoma smelter from the Quartz Creek mine. Prior to 1901, a small shipment of gold ore was made from the Carmack mine. Although the shipment was insignificant,

Barron gold camp (circa 1895)—typical of the many mining camps of that era

the Carmack mine is of historical interest because it was discovered by George Carmack, who is credited with the famous Klondike discovery of the north in 1896.

In Skagit County, rich deposits of lead and silver had been discovered at the headwaters of the Cascade River in 1889. The first discovery of the Cascade district was the Soldier Boy, which was followed by the discovery of other deposits that included the Boston, Midas, Chicago, and Johnsburg. Work was undertaken at many properties in the district, but only the Johnsburg became a producing mine. Small shipments were made in 1953 and 1955; the 1955 shipment of 9 tons, contained 57 percent lead and 40.4 ounces per ton in silver. By 1891, prospectors had moved north of the Cascade district and had discovered lead and silver at the headwaters of South Fork of Thunder Creek. High-grade lead ore from the Willis and Everett property contained as much as 3,400 ounces per ton in silver. Ore which contained 190 ounces per ton in silver, was shipped to the Everett smelter in 1893; however, high mining and shipping costs prevented further production at the mine.

In the Mount Baker district of Whatcom County significant discoveries of gold were made in 1897 and 1898. Both the Lone Jack and the Red Mountain mine became major gold producers with a combined production of around $1.5 million between 1902 and 1946. The Azurite mine in eastern Whatcom County also became one of the state's major gold producers. Although the Azurite was not discovered until 1915, it produced about $1 million in gold and silver between 1936 and 1939.

Elsewhere in western Washington deposits of gold, silver, and copper were discovered in the late 1800's and early 1900's in Pierce, Lewis, Skamania, and Clark Counties. Mining activity in these counties centered around the Silver Creek district in Pierce County, and the St. Helens district in Skamania County. Although hundreds of mining claims were staked and numerous deposits explored, only a few properties were placed into production, and the combined production of gold, silver, and copper from these four counties between 1903 and 1970 amounted to about $25,000.

By the early 1930's most of western Washington had been prospected, and nearly 40,000 mining claims had been staked. Sixty percent of the mining claims recorded in Washington were in Whatcom and Snohomish Counties, where significant discoveries of copper and gold had been made. Although work had been undertaken on hundreds of claims, no more than twenty-five properties became major producing mines. The most productive years for the mines were from 1890 until 1930; however, production was sporadic and most mines operated no more than 10 years. In order of decreasing value the mines produced gold, silver, and copper; the production of lead and zinc was insignificant. Only at the "45" mine in Snohomish County did the value of silver exceed that of the other metals in the ore. This made the "45" mine the only major silver mine in western Washington.

Production

Between 1903 and 1970 the bulk of the gold, silver, copper, lead, and zinc produced in western Washington came from Snohomish, Whatcom, and King Counties. Since 1962 no major metal mines have operated in western Washington. The combined metal production for western Washington counties from 1903 through 1970 totals $5,343,528; a breakdown by counties is shown in table 12.

The estimated silver production in western Washington from 1890 through 1969 is 2.25 million ounces. The leading silver-producing county was Snohomish, which produced 2,136,205 ounces; the

TABLE 12.—Metal production in western Washington, 1903-1970

County	Gold	Silver	Copper	Lead	Zinc	Total
Snohomish	$ 277,969	$199,487	$2,227,487	$ 437	...	$2,705,380
Whatcom	2,343,073	46,970	333	166	...	2,390,542
King	150,195	25,120	42,842	802	...	218,959
Pierce	486	356	6,448	206	...	7,496
Skagit	149	1,775	42	2,156	9	4,131
Lewis	60	69	1,760	...	...	1,889
Skamania	12,279	573	1,797	...	...	14,649
Clark	482	...	...	...	...	482
Total	$2,784,693	$274,350	$2,280,709	$3,767	$ 9	$5,343,528

bulk of the silver came from gold mines in the Monte Cristo district and from the Sunset copper mine near Index. Whatcom County is credited with around 80,000 ounces, most of which was a byproduct of gold mining operations in the Mount Baker district. King County produced around 36,500 ounces of silver, almost all of which was a byproduct of gold mining at the Apex mine. Lead-silver mines of Skagit County produced only around 2,380 ounces, whereas a total of less than 500 ounces of silver was produced by mines in Pierce, Lewis, Skamania, and Clark Counties.

TABLE 13.—Silver production of western Washington's major silver producing counties, 1903-1956

[*Prior to 1903 silver production was not reported on an individual county basis. After 1956 silver production is not reported to avoid disclosing confidential company information. In the table below, . . . denotes no production; c denotes production but it is concealed with other counties.*]

Year	County			
	Whatcom	Skagit	Snohomish	King
	Troy ounces			
1903	22,698	1,517	56,370	38
04	25,212	...	12,514	...
1905	22,111	...	29,946	c
06	465	...	5,084	c
07	244	...	19,389	535
08	71	...	53	1,732

Continued on next page

TABLE 13.—Silver production of western Washington's major silver producing counties, 1903-1956— Continued

Year	County			
	Whatcom	Skagit	Snohomish	King
	Troy ounces			
09	6	...	23	740
1910	...	4	32	3,145
11	1	...	5	3,499
12	24	...	6	2,778
13	95	...	673	3,038
14	188	...	1,371	...
1915	50	...	530	...
16	400	...	1,778	566
17	2,500	508	1,151	2,706
18	39	...	5,000	2,314
19	...	...	6,852	1,060
1920	3	...	14,648	693
21	91	...	2	...
22	247	...	1	360
23	164	...	10,018	...
24	38	...	8,646	...
1925	244	...	13,702	...
26	...	...	18,451	882
27	220	...	17,476	303
28	182	56	10,951	...
29	141	...	17,614	...
1930	173	...	9,277	...
31	35	...	1,350	...
32	246	...	7	...
33	23	...	14	...
34	116	...	17	116
1935	210	...	601	...
36	204	...	1,255	142
37	1,161	...	918	106
38	1,352	...	1,123	2,048
39	629	...	4,835	598
1940	121	...	9,187	322
41	180	...	16,591	1,139

Continued on next page

TABLE 13.—Silver production of western Washington's major silver producing counties, 1903–1956—Continued

| Year | County | | | |
	Whatcom	Skagit	Snohomish	King
	Troy ounces			
42	21	...	1,582	...
43	83	...	1,035	...
44	...	...	322	...
1945	...	...	218	...
46	...	...	349	...
47	115	...	...	...
48	...	...	105	31
49	21	...	1,106	...
1950	...	...	...	...
51	...	...	...	57
52	157	...	4	103
53	...	403	c	c
54	...	...	...	c
1955	...	125	c	c
1956	...	...	263	2,048
Total ounces	80,281	2,613	302,445	31,099

OCCURRENCES OF SILVER IN WASHINGTON

LOCATION OF DEPOSITS

In reviewing data on silver in Washington, the writer found it difficult to determine which properties should be covered by this report and which districts should be classed as silver districts. Silver is present in most of the state's base and precious metal deposits, but the average deposit contains no more than 1 ounce per ton. For the purpose of this report, a silver district is an area that contains several deposits from which silver has been mined, or an area containing deposits assaying 5 ounces or more per ton in silver. These districts are shown in figure 4. Eleven of the state's 20 silver districts are in the Okanogan Highlands physiographic province of northeastern Washington, and 9 districts are in the Cascade Mountains province.

GENERAL GEOLOGY

As in many mining districts of the west, most silver deposits in Washington appear to be related to granitic intrusive rocks. Although in several districts

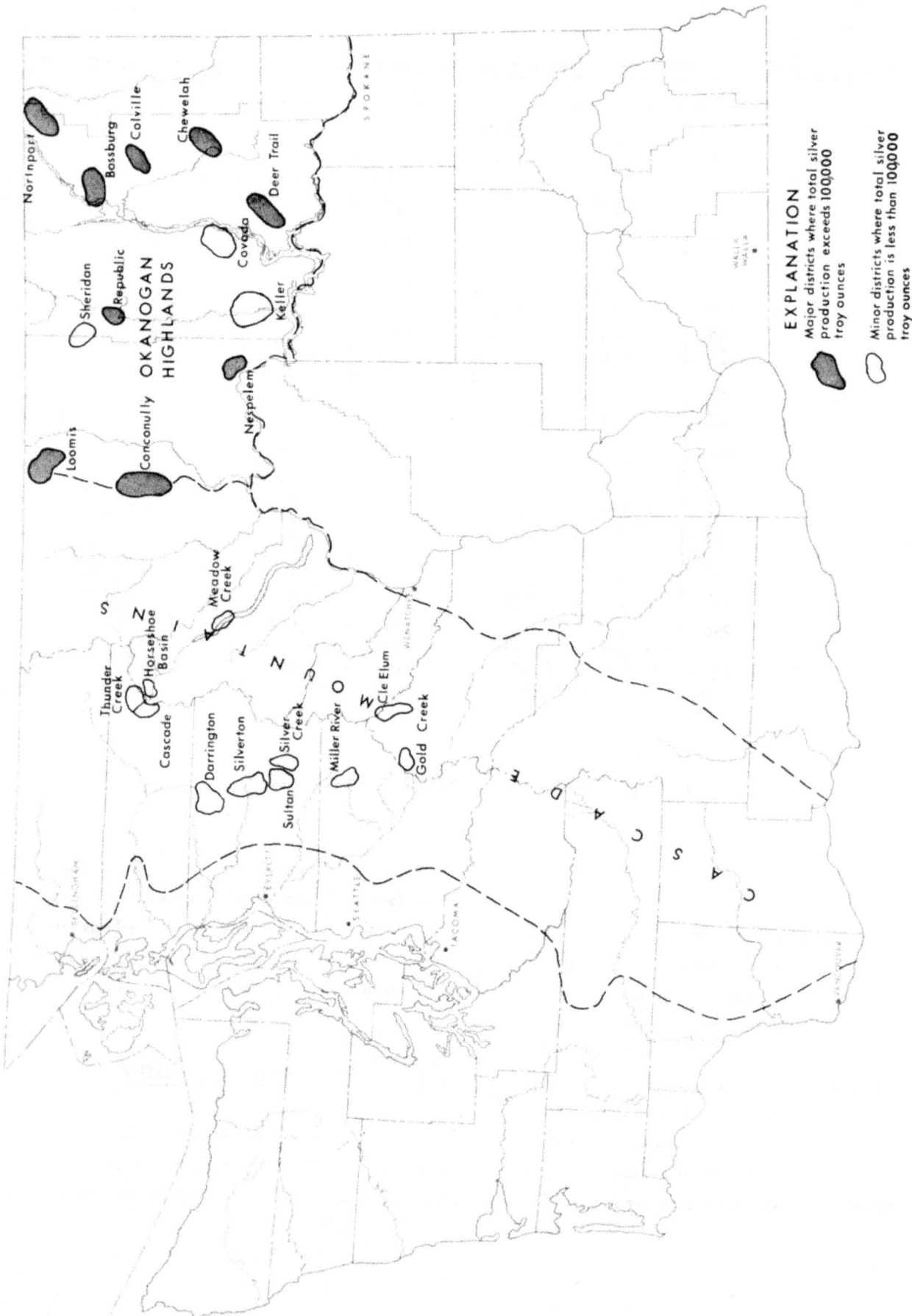

FIGURE 4.—Silver districts of Washington.

TABLE 14.—General location of silver districts in Washington

District	General location
Okanogan Highlands	
Northport	Northeastern Stevens County
Bossburg	Northwestern Stevens County
Colville	Central Stevens County
Chewelah	Southeastern Stevens County
Deer Trail	Southwestern Stevens County
Covada	Southeastern Ferry County
Republic	Northwestern Ferry County
Sheridan	Northwestern Okanogan and Northwestern Ferry Counties
Nespelem	Southeastern Okanogan County
Loomis	Northcentral Okanogan County
Conconully	Central Okanogan County
Cascade Mountains	
Cascade-Thunder Creek	Eastern Skagit County
Stehekin	Northwestern Chelan County
Darrington	North-central Snohomish County
Silverton	Central Snohomish County
Sultan	Southeastern Snohomish County
Silver Creek	Southeastern Snohomish County
Miller River	Northeastern King County
Snoqualmie	Northwestern Kittitas County
Cle Elum	Northwestern Kittitas County

the exposed granitic rocks may be a considerable distance from metal deposits, underlying granitic rocks are probably much closer. The general distribution of granitic rocks is shown in figure 5, which also shows the distribution of silver deposits, as well as base metal and gold deposits in Washington. It can be seen that silver deposits coincide with base metal and gold deposits.

In the Cascade Mountains and in the western part of the Okanogan Highlands argentiferous base metal deposits occur mainly in fissure veins and shear zones. The host rocks for the veins are chiefly lower to middle Tertiary quartz diorite and granodiorite and pre-Jurassic schist, gneiss, and hornfels. In the central part of the Okanogan Highlands, deposits of

gold, silver, copper, lead, and zinc occur in quartz fissure veins, breccias, and disseminated deposits. Common host rocks for the deposits consist of Tertiary volcanic rocks, early Tertiary and late Mesozoic granitic rocks, and Mesozoic and Paleozoic sedimentary rocks. In the eastern part of the Okanogan Highlands silver is associated with lead, zinc, copper, and gold, which is found in fissure veins, shear zones, and disseminated deposits. Common host rocks for the deposits are argillaceous and calcareous rocks of Cambrian age, as well as Cretaceous and Tertiary granitic intrusive rocks, consisting mainly of granodiorite. The silver-bearing deposits of the eastern part of the Okanogan Highlands fall within the Kootenay Arc, which is a narrow miogeosyncline of

FIGURE 5.—Distribution of silver in relation to the granitic rocks and deposits of gold, copper, lead, and zinc.

Granitic rocks   • Silver deposits

○ Gold, copper, lead, and zinc deposits

northeast-trending folded and faulted rocks. The Kootenay Arc is the dominant regional structure of Stevens and Pend Oreille Counties and adjacent to British Columbia. In Ferry County the richest silver deposits fall within the Republic graben, which is a major north-northeast-trending structural feature of western Ferry County. In Okanogan County and the Cascade Mountains, metal deposits appear to be related to north- to northwest-trending folds and faults, and transverse east-west-trending shear zones.

NATURE OF THE DEPOSITS

Most of Washington's silver deposits occur in narrow, steeply dipping quartz fissure veins and shear zones. The width of the veins ranges from thin stringers less than ½-inch in thickness to massive veins as much as 40 feet in thickness. Most veins have strike lengths of several hundred feet, whereas some veins can be followed along their strikes for several thousand feet. It is not uncommon for veins to pinch out abruptly along their strike and dip, and the vertical extent of most silver veins is not known. In the Conconully district of Okanogan County, some silver-bearing veins appear to extend to at least 2,500 feet beneath the surface. In the Republic district, the gold-silver veins have been mined to depths of 1,700 to 1,800 feet. However, based on old mine workings the average silver-bearing vein in Washington has not been mined for much more than 150 feet in depth.

The silver minerals are mostly sparsely disseminated in the veins, but in some veins they are concentrated into ore shoots. The average ore shoots are small and have strike and dip lengths of 10 feet or less. On the other hand, some ore shoots have been mined that had strike lengths of several hundred feet and dip lengths that were even greater. The veins may contain several ore shoots that are sporadically distributed in the vein. Individual ore shoots

consist of massive sulfides with little if any gangue, or disseminated ore minerals and lenses of ore minerals intermixed with gangue. In some veins the ore minerals occur in narrow black bands that parallel the walls. Seldom can the ore minerals be identified in the field because of their dustlike size. Post-mineral faulting is a prominent feature of many veins. The faulting has produced granulation of gangue and ore minerals, and in some cases has drawn out the ore minerals into thin filaments that parallel the faults. Transverse faults tend to offset some veins, and some veins cannot be traced beyond the faults.

In addition to ore shoots in quartz fissure veins, some silver deposits occur as irregular disseminated replacements along shear zones. These deposits assume the general shape of the sheared rocks and have a habit of pinching and swelling along the strike and dip of the deposit. Rather than a distinct quartz vein, the shear zone commonly contains brecciated country rock, siderite, calcite, and quartz. For the most part the ore minerals are disseminated in, or have replaced, parts of the shear zone. The ore shoots tend to be of an irregular, lenticular shape. Individual ore shoots range from 3 to 20 feet in thickness and have strike and dip lengths of up to several hundred feet. Although some shear zones are metallized to depths of 1,000 feet beneath their outcrops, the richest parts of the shear zones tend to be within 100 feet of the surface.

The chief gangue minerals of the vein deposits are quartz, calcite, siderite, and dolomite; barite is a less common gangue mineral. Except in oxidized parts of the vein, pyrite, pyrrohotite, or arsenopyrite may also be present. In breccia and replacement deposits in limestone or dolomite, the principal gangue mineral is massive quartz that is often accompanied by calcite, siderite, dolomite, and pyrite.

Most silver veins in Washington are only slightly oxidized, and extensive downward enrichment is lacking. In the Nespelem district of Okanogan County, the oxidized zone extends to around 40 feet beneath the surface, and in the Republic district it may be as great as 150 feet. Oxidation in both districts has not been complete and sulfide minerals in some veins occur almost at the surface. Available data indicate that most oxidized zones of silver deposits in the state are impoverished rather than enriched. In the near-surface and central parts of oxidized zones, some silver occurs as native silver in the form of leaves, small flakes, arborescent masses, and tangled wires. Parts of the veins that occur near water level are sometimes enriched by argentite, pyrargyrite, cerargyrite, and native silver. In almost all oxidized parts of the veins, iron and manganese oxides are common.

ORIGIN AND AGE

Almost all investigators of silver, gold, copper, lead, and zinc deposits in Washington attribute the origin of the metals to deep-seated granitic rocks. During cooling phases of granitic rocks, siliceous hydrothermal solutions rich in metals tend to segregate within parts of the congealing magma. When fractures develop within the rock that retains the hydrothermal solutions, the solutions ascend toward the surface along fractures in the earth's crust. Near the surface, temperatures and pressures drop and minerals precipitate out to form veins in open fissures, or to replace other minerals in the wall rock of fractures and fissures, forming replacement deposits. Recurrent fracturing of pre-existing veins, along with the introduction of other metalliferous solutions, can produce several generations of metals. Although most silver deposits in Washington appear to have originated during one period of metallization, the mineralogy of

some deposits suggests at least two separate periods, and possibly more.

The age of most silver deposits in Washington appears to be Late Cretaceous-early Tertiary, as are most of the intrusive granitic rocks of the state. An exception to this is the gold-silver deposits of the Republic district, which are of Oligocene-Miocene age.

MINERALOGY

In Washington silver minerals are most commonly associated with the ore minerals of lead, copper, and gold, and less commonly with zinc and antimony minerals. In deposits where silver minerals predominate, the common silver minerals are argentite, stephanite, cerargyrite, proustite, pyrargyrite, and native silver. Associated ore minerals consist chiefly of galena, sphalerite, chalcopyrite, bornite, tetrahedrite, pyrite, pyrrhotite, and arsenopyrite. In lead deposits, where galena is the main ore mineral, silver is contained in argentite, tetrahedrite, and tennantite, that occur as minute inclusions in the galena. In some deposits one or more of these silver minerals occur interstitial to grains of galena. In copper deposits the silver is usually contained in minute grains of argentiferous tetrahedrite, which coat grains of chalcopyrite and(or) bornite. The tetrahedrite may also occur interstitial to copper minerals, and parts of some copper veins have consisted almost entirely of tetrahedrite. Some gold deposits contain silver minerals in addition to the silver that is almost always alloyed with the gold. Argentian tetrahedrite, argentite, pyrargyrite, native silver, sylvanite, and naumannite occur in several of the state's gold deposits. One or several of these silver minerals may occur as sparsely scattered grains in the gold veins, or in narrow black streaks, which represent finely granulated minerals that rarely can be identified. Argentian tetrahedrite appears to be the most common silver mineral of gold veins. Silver is also present in several of the state's antimony deposits, but very few silver minerals have been identified. In most deposits the silver content of the ore may be attributed to argentiferous galena and tetrahedrite; however, at the Wells Fargo mine in Stevens County, where the silver content of the ore is as much as 25 ounces per ton, silver occurs in argentiferous zinkenite.

To date (1975) the bulk of silver mined in the state has come from gold, copper, and lead ore. Lead-zinc ores of Stevens and Pend Oreille Counties contain as little as 0.02 ounce per ton in silver, whereas high-grade lead ore mined in these counties has contained as much as 1,000 ounces. Copper ore from the Chewelah district of Stevens County contained up to 100 ounces, but the average silver content of mined ore was only 4 ounces per ton. At the Sunset mine in Snohomish County, the average silver content of high-grade copper ore was only 0.59 ounce per ton. All gold ore contains silver, usually in the form of gold-silver alloy. In some parts of the Republic district of Ferry County, the gold-silver ratio of the ore is as high as 1:12, whereas the average for the district is about 1:4. At other gold mines in the state gold ore contained as little as 0.5 ounce of silver for each ounce of gold. In deposits that contain mainly silver, and only minor base metals and gold, the average silver content of mined ore was around 60 ounces per ton. Much near-surface ore ran several hundred ounces per ton in silver, while a few high-grade shipments from several of the state's silver districts ran as high as 1,000 ounces per ton.

Of 63 known silver minerals, 15 have been reported in Washington. The most common silver min-erals in the state include tetrahedrite, argentite, pyrargyrite, stephanite, and native silver in order of decreasing abundance. A brief description of these and other silver minerals of the state follows:

Andorite - $PbAgSb_3S_6$
A dark-gray to black silver mineral that occurs as prismatic crystals. A comparatively rare mineral.

Argentite - Ag_2S
Most common silver mineral and generally of hypogene origin. Dark lead gray, which tarnishes on surface to a black earthy sulfide. Occurs in large masses and as microscopic inclusions in galena and other sulfide minerals.

Bromyrite - $AgBr$
Bright-yellow to amber yellow, slightly greenish. Occurs in oxidized zones of silver deposits along with native silver and other supergene minerals.

Cerargyrite - $AgCl$
Highly sectile, usually massive and re-sembles wax. Known as horn silver, it often occurs in crusts. Pearl gray to colorless but becomes violet to brown on exposure to light. Secondary mineral in upper parts of silver deposits.

Diaphorite - $Pb_2Ag_3Sb_3S_8$
Light steel gray to silver white, also blackish lead gray. A comparatively rare mineral.

Dyscrasite - Ag_3Sb
Silver white, usually tarnished to lead gray; sometimes yellow or blackish. Associated with galena, ruby silver, argentite, and native silver.

Native silver - Ag

Silver white, tarnishes to gray or black. Ductile and malleable. May be primary in origin, but usually is secondary and found in upper part of silver deposits. Commonly occurs in arborescent and wiry forms, irregular masses, and scales. Nearly always present in native gold.

Naumannite - Ag_2Se

A comparatively rare mineral of hypogene origin. Iron black and occurs in cubic crystals and in thin plates, as well as massive and granular.

Proustite - $3Ag_2S-As_2S_3$

A ruby silver mineral that is scarlet vermilion, brittle, and transparent to translucent. A hypogene mineral in upper parts of silver veins.

Pyrargyrite - Ag_3SbS_3

Ruby silver mineral so called because it is dark red in thin splinters. Commonly black to grayish black and brittle. Common hypogene mineral in upper parts of silver veins.

Stephanite - Ag_5SbS_4

Brittle, iron black, and commonly associated with other silver minerals in upper parts of veins. Hypogene in origin.

Stromeyerite - $(Ag,Cu)_2S$

A fairly common silver mineral, especially in deposits that contain high contents of copper. Dark gray and tarnishes blue on exposed surfaces. May be hypogene or supergene, and the latter is often associated with chalcocite.

Tetrahedrite - $(Cu,Fe,Zn,Ag)_{12}(Sb,As)_4S_{13}$

High silver varieties are called argentian tetrahedrite, argentian tennantite, and freibergite. Flint gray to iron black, and rather brittle. Generally a hypogene mineral that is commonly associated with chalcopyrite, bornite, pyrite, sphalerite, galena, argentite, and pyrargyrite.

PART II

SILVER OCCURRENCES
of
NORTHEASTERN WASHINGTON

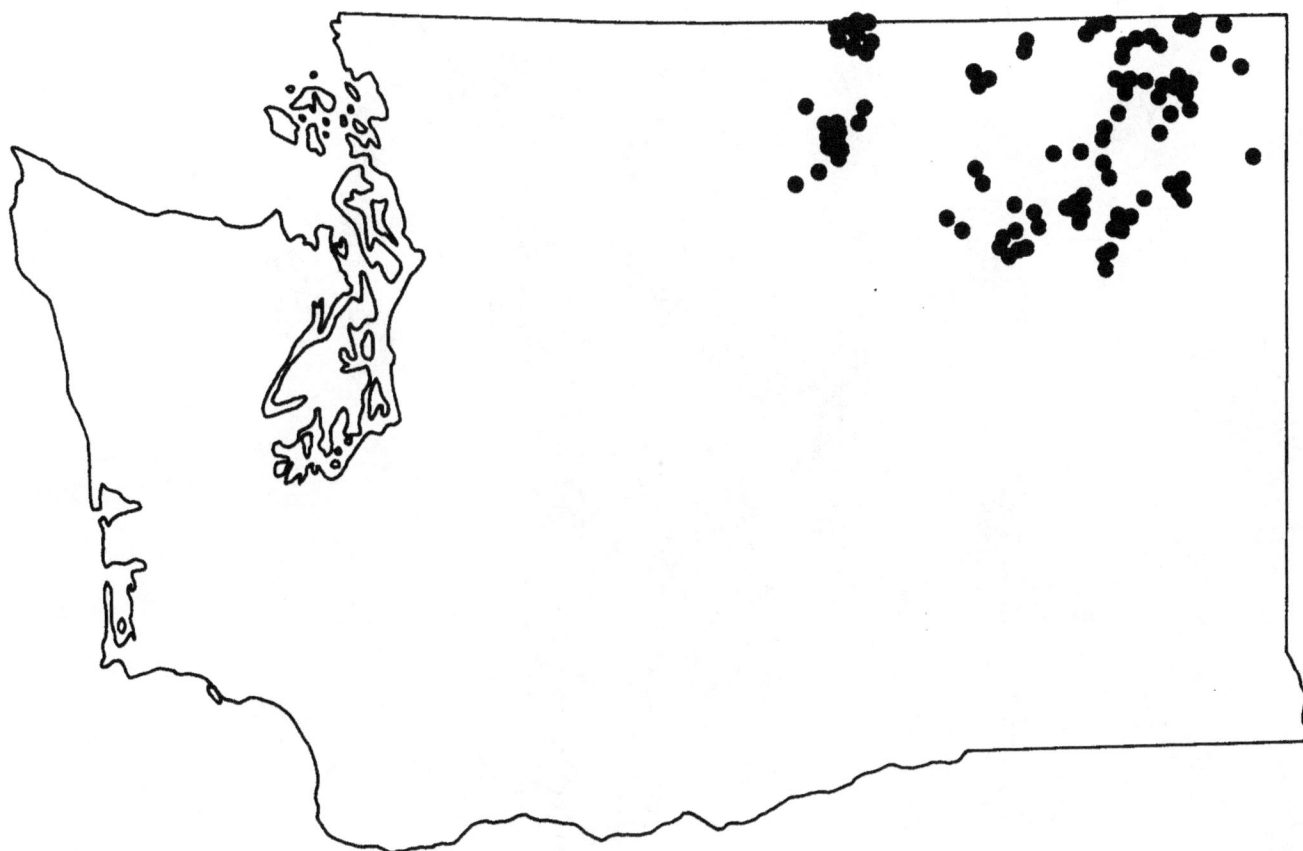

OKANOGAN HIGHLANDS

INTRODUCTION

Most of the state's silver deposits and the more important silver districts are in the Okanogan Highlands physiographic province of northeastern Washington. Silver has been reported at 703 properties and is a metal of principal value at 197 properties. To date (1975) over 200 mines have produced silver; however, most silver was a byproduct of lead, copper, and gold mining operations.

Although silver is widespread throughout northeastern Washington, the most significant deposits, as well as the major silver mines that have produced in the past, fall within 11 mining districts. These districts are Loomis, Conconully, Nespelem, and Sheridan, in Okanogan County; Republic and Covada, in Ferry County; and Northport, Bossburg, Colville, Chewelah, and Deer Trail, in Stevens County. Stevens County leads in the number of mines and prospects containing silver, followed in decreasing order by Okanogan, Ferry, and Pend Oreille Counties. A breakdown of silver occurrences by county and the principal metal at each property as reported by Huntting (1956) is listed in table 15.

PHYSIOGRAPHY

The Okanogan Highlands are bordered on the west by the Cascade Mountains and on the south by the Columbia Basin; they extend northward into Canada and eastward into Idaho. The highlands are

TABLE 15.—Distribution of silver in Pend Oreille, Stevens, Ferry, and Okanogan Counties
(Mines and prospects where silver occurs)

Principal metal at mine or prospect	Counties				No. of mines and prospects
	Pend Oreille	Stevens	Ferry	Okanogan	
Silver	15	47	49	86	197
Lead	13	91	56	37	197
Gold (lode)	0	35	52	57	144
Copper	11	56	16	28	111
Zinc	17	17	2	4	40
Iron	0	1	1	2	4
Antimony	0	1	2	0	3
Molybdenum	0	2	0	1	3
Nickel	0	0	2	1	3
Tungsten	0	1	0	0	1
Total	56	251	180	216	703

mountainous, but they lack the ruggedness of the Cascade Mountains. East of the Columbia River are the Huckleberry, Chewelah, and Selkirk Mountains; west of the river are the Kettle, San Poil, and other ranges. General summit elevations in the highlands range from 3,000 to 5,000 feet; however, rocky peaks rise above these levels, and on several peaks elevations exceed 7,000 feet. Most mountain slopes and summits are thickly wooded, but in some parts of the highlands sparsely wooded, steep, rocky hillsides and peaks dominate the local landscape. Numerous streams have thoroughly dissected the Okanogan Highlands. All streams ultimately drain into the Columbia River, which is the region's major drainage. Other major rivers are the Colville, Pend Oreille, San Poil, Kettle, and Okanogan. Elevations along the rivers range from lows of 800 to 900 feet along the Okanogan to highs of 2,000 to 2,100 feet along the Pend Oreille.

GENERAL GEOLOGY

As can be seen in figure 6, the Okanogan Highlands contain a wide variety of rocks that vary in age from Precambrian to Holocene. Throughout the Highlands, most major valleys and half the mountain slopes are covered by deposits of silt, sand, and gravel of glacial origin.

West of the Columbia River, in Ferry and Okanogan Counties, Mesozoic and early Tertiary granitic rocks of the Colville and Similkameen batholiths predominate. The granitic rocks, consisting mainly of granodiorite, intrude pre-Jurassic schist, quartzite, phyllite, greenstone, and amphibolite in northeastern Ferry County and north-central Okanogan County. In southeastern Ferry County, the granitic rocks intrude Paleozoic metasedimentary rocks, con-

sisting chiefly of graywacke, impure quartzite, slate, greenstone, and limestone. Near the central part of the Okanogan Highlands, the Republic graben, an elongated down-faulted block 55 miles long and about 15 miles wide is filled with great thicknesses of Tertiary andesitic lavas and flow breccias, as well as conglomerate and siltstone. In the southwestern corner of the highlands, Columbia River basalts of Miocene age have been deposited upon erosion surfaces of Mesozoic granitic rocks. These basalts are the northern limits of lava flows covering most of southeastern Washington.

East of the Columbia River in Stevens and Pend Oreille Counties, Precambrian and Paleozoic metamorphic and sedimentary rocks predominate. Precambrian rocks are mainly quartzite, argillite, and impure dolomite; Paleozoic rocks consist mainly of quartzite, phyllite, limestone, and dolomite. The limestone and dolomite are important host rocks for silver-lead deposits in Stevens County and lead-zinc deposits in Pend Oreille County. Throughout both counties the Precambrian and Paleozoic rocks have been intruded by Mesozoic granodiorite of the Loon Lake batholith.

In the eastern Okanogan Highlands, northeast-trending folds predominate. They are part of the Kootenay Arc—a northeast-trending miosyncline formed during mountain-building activity that began in the Middle Jurassic and continued into early Tertiary. As the rocks were folded, northwest- and northeast-trending faults developed. In the western Okanogan Highlands, northwest-trending folds and faults predominate and conform with northwest-trending structures of the Northern Cascades. In the valley of the Columbia River between Ferry and Stevens Counties, folds trend both northeast and northwest with some intermediate north trends.

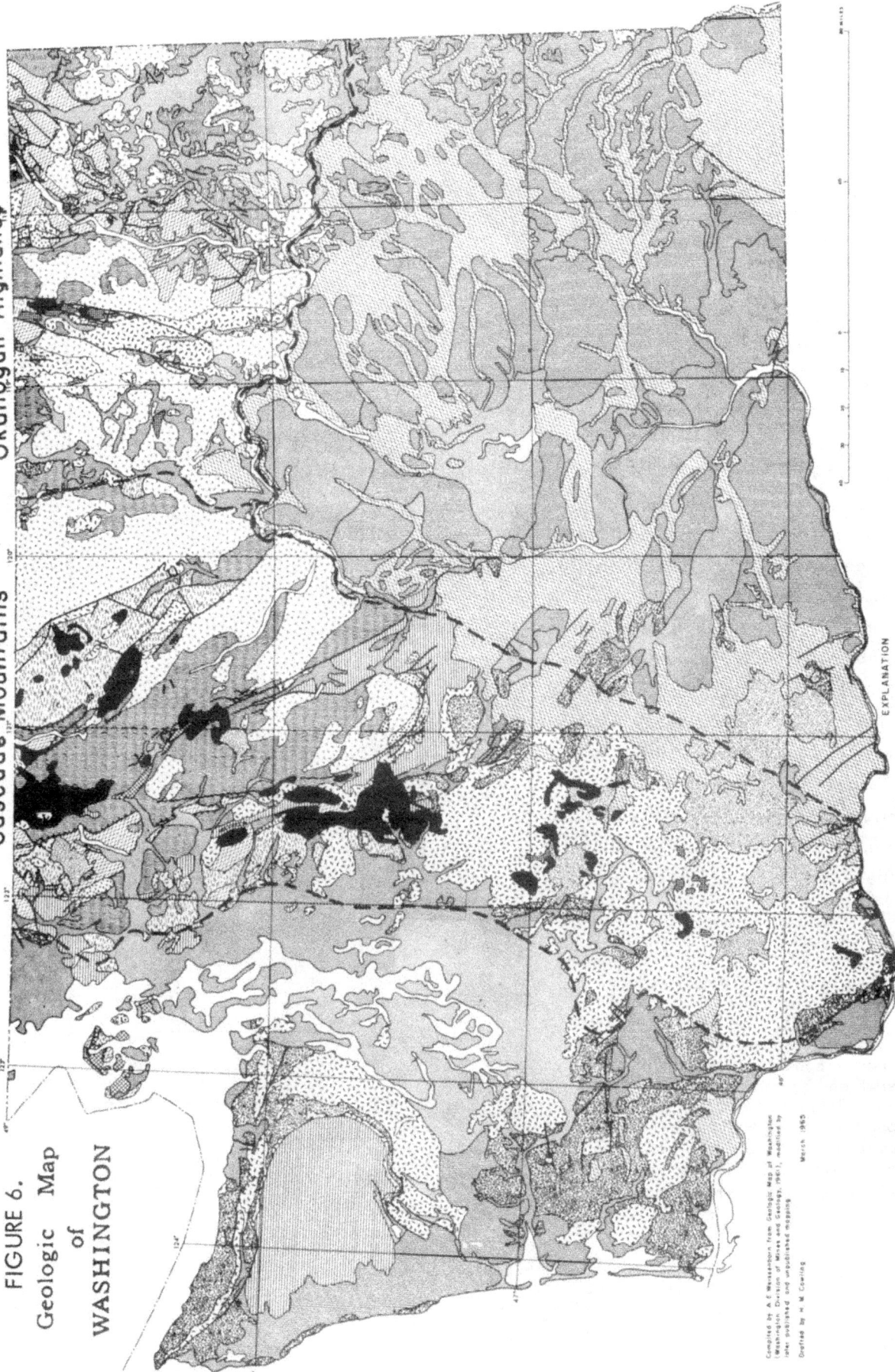

FIGURE 6.
Geologic   Map
of
WASHINGTON

PEND OREILLE COUNTY

Although there never have been any major silver mines in Pend Oreille County, silver has been an important byproduct of lead-zinc mining operations. From 1906 through 1969, 30 mines in the county produced 708,330 ounces of silver from ores that averaged only 0.040 ounce per ton in silver. In addition to silver the mines produced 427,041 tons of zinc, 196,373 tons of lead, 526 tons of copper, and 266 ounces of gold. The combined value of the silver, gold, copper, lead, and zinc totaled $145,944,035.

Currently (1975), the Pend Oreille mine in the Metaline district is the county's only producing metal mine; it produces around 6,000 ounces of silver, 10,000 pounds of copper, 4 million pounds of lead, and 13 million pounds of zinc yearly.

The bulk of the silver produced in Pend Oreille County has come from the Metaline district; less than 1 percent has come from the Newport district, which is the only other mining district in the county. The silver-bearing deposits of the Metaline district occur chiefly in the "Josephine Horizon" of the Metaline Limestone (Cambrian). Addie (1970, p. 75) describes the ore bodies of the "Josephine Horizon" as follows:

> The ore bodies range from pods and lenses a few feet thick and 10 to a few tens of feet wide and long, to masses 3,000 feet long, 100 feet thick, and 300 feet wide. The ore bodies are highly irregular, and many are elongated in channel-like shapes that pinch and swell, and coalesce irregularly with nearby ore pods. The longest elongated bodies plunge northeastward at 10° to 20°.
>
> The ore consists of an intimate mixture of sphalerite and galena in a wide range of proportions. This range is sufficient that certain parts of the ore bodies can be characterized as lead-rich areas, others as zinc-rich areas. Silver minerals have not been identified. Pyrite is the common gangue

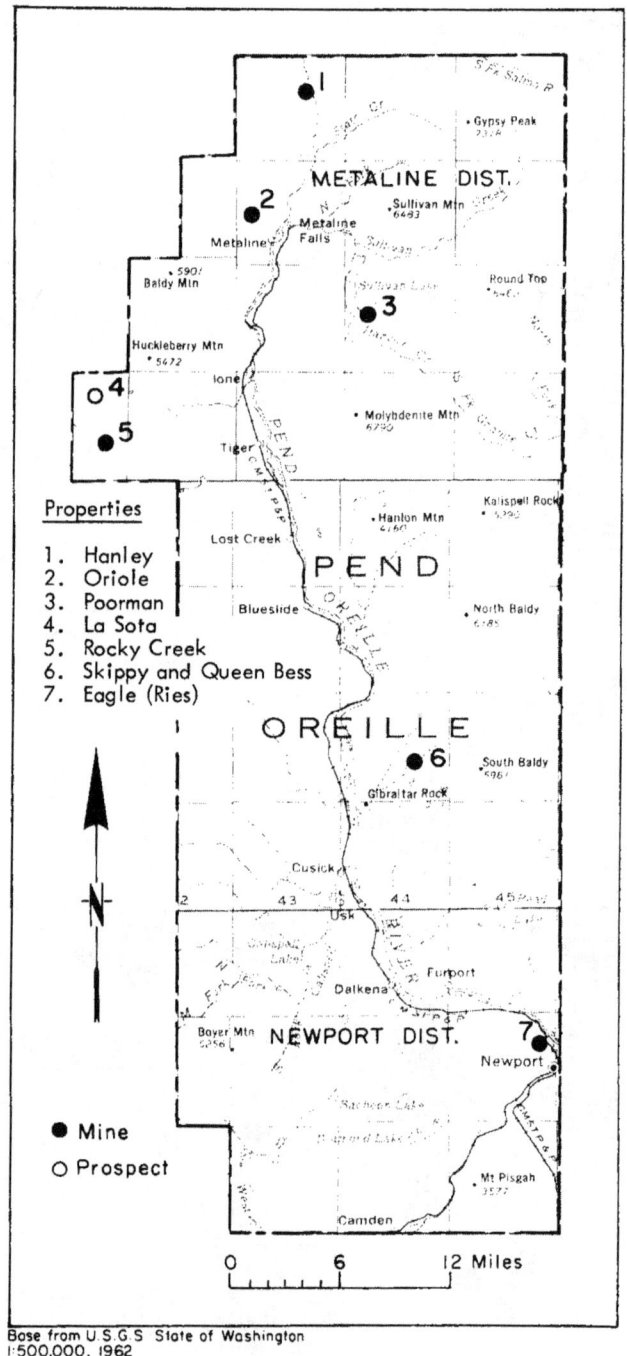

FIGURE 7.—Silver deposits of Pend Oreille County.

sulfide. Other gangue minerals are quartz, calcite, dolomite, barite, and palygorskite.

Sphalerite occurs chiefly as fine disseminations in the matrix of the breccias, and also rims breccia fragments. In "zebra rock," sphalerite may form narrow bands along one edge of white dolomite streaks. Coarser grained sphalerite, with quartz, forms irregular vein-like streaks and patches in the matrix and between breccia fragments. It may occur also within white, coarse-grained dolomite fragments that are fragments of the breccia. Sphalerite also replaces fine-grained gray limestone, either as coarse-grained aggregates or as fine-grained masses in which individual grains cannot be detected in hand specimens. The color of sphalerite varies from reddish brown to pale yellow.

Most of the galena is distributed irregularly throughout the breccia as stringers, pods, and segregations. It is ordinarily medium grained, but large cubes are found occasionally. In places, it fills faults and fractures in the breccia matrix.

Small, irregular bodies and scattered grains of sphalerite and galena are found throughout the entire stratigraphic thickness of the Metaline Limestone, except in the very basal part. All the presently mined ore bodies, however, are in the upper 200 feet of the Metaline Limestone. Most ore bodies are found in the breccia. Most of the breccia contains at least traces of sphalerite and galena, but not all of it is ore. Very few ore bodies are in contact with the Ledbetter Slate.

A small tonnage of more pyritic ore was mined some years ago from a second zone, known locally as the "Yellowhead Horizon," 1,100 to 1,200 feet stratigraphically below the Ledbetter Slate. This zone is exposed in only a few places and has been little explored in the Metaline district. It has been a productive ore zone to the west in parts of Stevens County, Washington. [In 1970, diamond drilling disclosed significant lead-zinc stratiform mineralization in the "Yellowhead Horizon" and subsequent drilling on the ore body encountered milling ore.]

In the Metaline district no ore bodies have been found in the Maitlen Phyllite, which underlies the Metaline Limestone, although important ore bodies are found in carbonate rocks in Stevens County, Washington, and in its Canadian equivalent, the Laib Formation in adjoining parts of British Columbia. Minor amounts of ore have been found at the Oriole mine in a limy zone near the base of the Gypsy Quartzite, which strati-graphically underlies the Maitlen. The Oriole is west of the graben outside the main producing area of the district.

In addition to the silver-bearing lead-zinc deposits in the Metaline Limestone of the Metaline Falls area, silver is present elsewhere in the county, but occurs chiefly in metallized quartz veins and shear zones. These deposits are found in a variety of host rocks which include limestone, dolomite, argillite, phyllite, quartzite, and greenstone. The quartz veins and shear zones are from 1 to 6 feet wide, and carry pyrite, chalcopyrite, tetrahedrite, galena, sphalerite, and, rarely, native silver. The ore minerals are generally sparse, but in a few deposits they are concentrated into small ore shoots. To date (1975), less than 3,000 tons of ore has been shipped from the deposits. Although the ore averaged only around 10 ounces per ton in silver, small shipments have yielded as much as 1,190 ounces of silver per ton. As many as 56 properties in Pend Oreille County have reported the presence of silver, but only at the following properties has it been reported to be present in significant amounts.

METALINE DISTRICT

Principal Silver Deposits

Hanley

Location: NE¼ sec. 10, T. 40 N., R. 43 E.

Development: 20-foot shaft and large open cut.

Geology: Replacement deposit in silicified limestone, traceable for 300 feet on surface.

Ore minerals: Argentiferous galena.

Production: 2 carloads of high-grade lead-
　　silver ore in the 1880's.
Reference: Huntting, 1956, p. 368.

LaSota

Location: SE$\frac{1}{4}$ sec. 3 and NE$\frac{1}{4}$ sec. 10,
　　T. 37 N., R. 41 E.
Development: 40-foot shaft, and 3 adits
　　which total 350 feet in length.
Geology: 1$\frac{1}{2}$- to 6-foot-thick quartz veins
　　in phyllite and schistose greenstone
　　assay up to 11.5 ozs. per ton in silver.
Ore minerals: Galena, sphalerite, bornite,
　　chalcopyrite, and tetrahedrite.
Production: None.
Reference: Huntting, 1956, p. 313.

Oriole

Location: SE. cor. sec. 19, T. 39 N.,
　　R. 43 E.
Development: 135-foot shaft.
Geology: Quartz lenses up to 3 feet thick
　　and 20 feet long in brecciated dolo-
　　mite. Hand-sorted ore contained
　　42 ozs. silver, 22 percent zinc, 15
　　percent lead, and 1 percent copper.
　　Ore averaged 5.8 ozs. per ton in
　　silver.
Ore minerals: Sphalerite, galena, tetrahed-
　　rite, chalcopyrite, pyrite, smithso-
　　nite, cerussite, and bornite.
Production: 2,000 tons prior to 1943.
Reference: Huntting, 1956, p. 371.

Poorman

Location: Center sec. 17, T. 38 N.,
　　R. 44 E.
Development: Unknown.
Geology: Unknown.

Ore minerals: Unknown
Production: 2 tons of ore mined in 1937
　　yielded 2,380 ozs. silver, 14 lbs.
　　copper, and 18 lbs. lead.
Reference: Fulkerson and Kingston, 1958,
　　p. 23.

Rocky Creek

Location: NE. cor. sec. 26 and SW$\frac{1}{4}$ sec.
　　23, T. 37 N., R. 41 E.
Development: Caved shaft and two adits
　　each several hundred feet in length.
Geology: 2-foot-thick quartz vein in argil-
　　lite and quartzite. Parts of the vein
　　contained 360 ozs. per ton in silver.
　　Most of the vein averages less than 1
　　oz. per ton in silver.
Ore minerals: Galena, sphalerite, chalco-
　　pyrite, and native silver.
Production: 1953, 1956, 1958, and 1959:
　　9 tons yielded 3 ozs. gold, 1,238
　　ozs. silver, 1,240 lbs. lead, and
　　1,715 lbs. zinc; 275 tons yielded 5
　　ozs. gold, 263 ozs. silver, 300 lbs.
　　copper, 9,400 lbs. lead, and 6,700
　　lbs. zinc.
Reference: Huntting, 1956, p. 314.

NEWPORT DISTRICT

Principal Silver Deposits

Ries (Eagle)

Location: N$\frac{1}{2}$NW$\frac{1}{4}$ sec. 12, T. 31 N.,
　　R. 45 E.
Development: 80- and 215-foot shafts, and
　　a 40-foot adit.

Geology: 6-foot-thick quartz vein in quartz
diorite porphyry.

Ore minerals: Galena and minor chalcopyrite.

Production: 1910-1922: 21 tons of ore
yielded 5 ozs. gold, 363 ozs. silver,
414 lbs. copper, and 973 lbs. lead.

Reference: Huntting, 1956, p. 229.

Skippy and Queen Bess

Location: Sec. 23 and SW¼ sec. 14,
T. 34 N., R. 44 E.

Development: 80- and 105-foot adits.

Geology: 10- to 12-inch-thick quartz vein
in quartzite.

Ore minerals: Galena and pyrite.

Production: None.

References: Huntting, 1956, p. 229;
Schroeder, 1952, p. 53-54.

STEVENS COUNTY

Stevens County ranks second in the production
of silver in Washington, having produced 3,378,763
ounces of the metal from 1904 through 1969 (table 5).
Metal deposits containing silver occur in many parts
of the county, but the bulk of the silver has come
from five of the county's nine mining districts. These

TABLE 16.—Major silver-producing districts of Stevens County, 1902-1969

District	Predominant metals	Total silver production	Average silver content (oz/ton) of ore
Chewelah	Copper-silver	1,792,219	4.5
Colville	Lead-silver	366,717	53.68
Bossburg	Lead-silver	307,288	2.62
Northport	Zinc-lead	239,956	0.075
Deer Trail	Lead-silver	190,602	9.82

TABLE 17.—Major silver producers of Stevens County, 1902-1956

Mine	District	Chief metal	Years operated[1]	Production (troy ozs)
United Copper	Chewelah	Copper	1906-1931	1,645,997
Old Dominion	Colville	Lead	1902-1953	342,517
Queen Seal	Deer Trail	Silver	1903-1940	285,759
Bonanza	Bossburg	Lead	1907-1952	238,485
Legal Tender	Deer Trail	Silver	1902-1904	122,211
Cleveland	Deer Trail	Lead	1903-1948	98,745
Providence	Deer Trail	Silver	1902-1911	87,442
Young America	Bossburg	Lead	1905-1953	69,893

[1] Operations not continuous.

Properties

1. Comstock
2. Iron Mask
3. Montana and Washington
4. Easter Sunday
5. McKinley
6. Bullion
7. Great Republic
8. Coyote
9. Sunset
10. Sterret
11. Silver Crown
12. Melrose
13. Jackson
14. Red Top
15. Hazel
16. Roosevelt
17. United Treasure
18. Keough
19. Myeerah
20. Frisco Standard
21. Morning
22. Burrus
23. Galena Farm
24. Al Ki
25. Silver Trail
26. Young America
27. Bonanza
28. Gold Bar
29. Chloride Queen
30. Avondale–Dome
31. Gold Ledge
32. Vanasse
33. Gold Reef
34. Ark
35. Aguila
36. Longshot
37. Middleport
38. Old Dominion
39. Ore Cache
40. Daisy Tempest
41. Silver Summit
42. Krug (Hanford)
43. Jay Dee
44. High Grade
45. United Copper
46. Copper King
47. Amazon
48. Chinto
49. Eagle (Blue Star)
50. Jay Gould
51. U.S. Copper Gold
52. Mullen
53. Nevada
54. Edna
55. Double Eagle
56. Wells Fargo
57. Orchid
58. Cleveland
59. Deer Trail
60. Brooks
61. Saturday Night–
    Sunday Morning
62. Queen-Seal
63. Aichan Bee
64. Silver Star
65. Orazada
66. Indian Trail

● Mine

○ Prospect

0   6   12 Miles

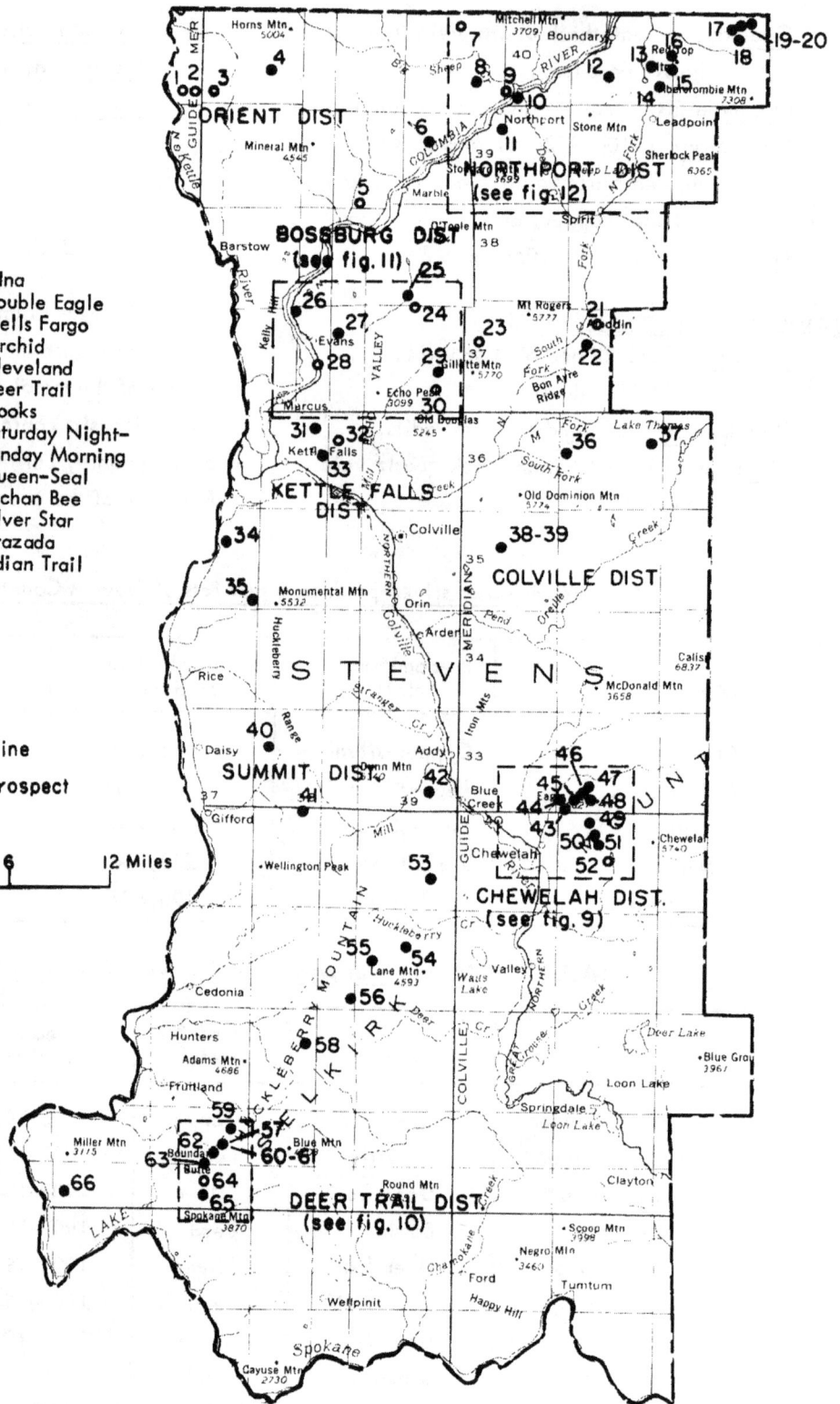

FIGURE 8.—Silver deposits of Stevens County.

five districts, along with predominant metals of the districts, silver production, and average silver content of the ore, are listed in table 16.

According to U.S. Bureau of Mines figures (Fulkerson and Kingston, 1958, p. 24-46), 193 mines in Stevens County are credited with the production of silver from 1902 through 1956. However, only at 38 mines did the value of silver exceed that of any other single metal, and only at 8 mines did the total silver production exceed 50,000 ounces. A breakdown, according to the main metal produced, shows that silver was produced by 70 lead mines, 38 silver mines, 37 gold mines, 32 copper mines, and 16 zinc mines. Mines that produced in excess of 50,000 ounces of silver are listed in table 17.

CHEWELAH DISTRICT

Location

The Chewelah district of southeastern Stevens County is one of the oldest mining districts in Washington. In 1883, lead and silver deposits were discovered at Embry camp, which was 2 miles east of Chewelah. Subsequent discoveries proved to be rich in copper and silver, which lead to the development of some of the most productive silver mines in the state. Although silver occurs throughout the district, the most productive mines are 3 to 5 miles east of Chewelah near Eagle Mountain and Quartzite Mountain (fig. 9). Several other silver mines are 12 miles

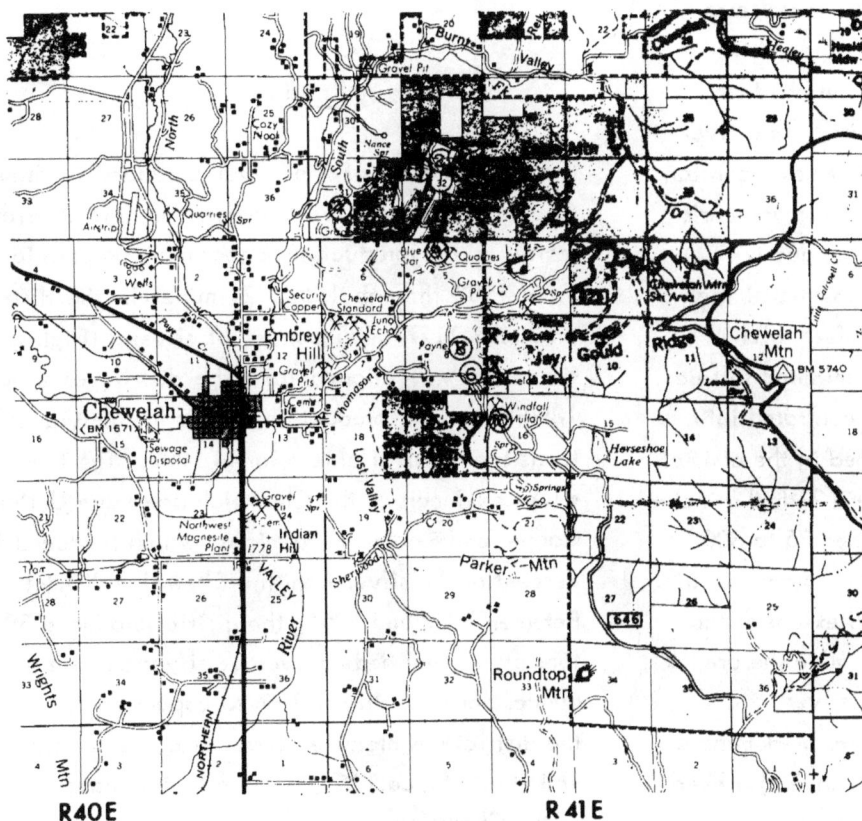

Mines and Prospects

1. Chinto
2. Copper King
3. Amazon
4. High Grade
5. United Copper
6. U.S. Copper Gold
7. Jay Dee
8. Jay Gould
9. Eagle (Blue Star)
10. Mullen

Base map U.S.F.S.

FIGURE 9.—Index map of the Chewelah district.

southwest of Chewelah, but production from these mines has been minor. Currently (1975), no metal mines are operating in the district.

Geology and Mineralization

The predominant rocks of the Chewelah district consist of interbedded argillite, slate, phyllite, and quartzite of the Belt Series (Precambrian); argillite, quartzite, phyllite, and dolomite of the Deer Trail Group (Precambrian); Addy Quartzite (Cambrian); and Flowery Trail Granodiorite (Mesozoic). Mafic dikes occur throughout the district in metamorphic and plutonic rocks. High-angle, northeast-trending faults and shear zones served as depositional loci for metal deposits, consisting of metallized veins of white quartz and carbonate minerals (calcite to siderite). The chief sulfide minerals of the veins are pyrite, pyrrhotite, chalcopyrite, tetrahedrite, covellite, chalcocite, and molybdenite. The silver occurs chiefly in the tetrahedrite, either as argentian tetrahedrite or as freibergite. Other silver minerals appear to be absent in the district. The quartz-carbonate veins range from stringers less than 1 inch thick to massive veins as much as 25 feet thick. Most veins contain only sparsely disseminated sulfides, but in some veins the sulfides are concentrated into ore shoots. The richest ore shoot mined in the district was at the United Copper mine where a 2- to 12-inch-thick ore shoot of tetrahedrite contained 75 to 300 ounces of silver per ton. This shoot had a horizontal length of 100 to 200 feet; the vertical extent of the shoot is not known. Elsewhere in the mine the ore averaged only 4.63 ounces per ton in silver.

Although quartz-carbonate veins predominate in the Chewelah district, ore minerals have also been found in chimneylike breccia zones in limestone. At the Eagle (Blue Star) mine, argentiferous galena and cerussite occurred in a chimney of brecciated limestone that was several hundred feet in diameter. However, the lead-silver ore bodies of the breccia zone were sparse.

Near-surface parts of most veins contain iron and manganese oxides, while oxidized parts of the copper veins contain malachite and azurite. No large zones of secondary enrichment occur in the district; if they did exist, they were probably removed by glacial erosion.

Most silver-bearing metal deposits of the Chewelah district occur in metamorphic rocks that border granodiorite intrusives. As such, the ore deposits appear to have originated from hydrothermal metalliferous solutions that formed vein-type deposits in fissures, while other solutions formed metosomatic replacement deposits along shear zones.

Production

The bulk of the silver produced in Stevens County came from the Chewelah mining district, mainly as a byproduct of copper mining operations. From 1902 through 1957, 24 mines in the district produced 1,771,435 ounces of silver. High-grade silver ore was shipped from the district as early as 1890; however, production figures for individual districts are not available prior to 1902. The leading silver producers of the Chewelah ores were United Copper and Copper King mines, which produced 95 percent of the silver from the Chewelah district. Between 1902 and 1956, the district produced 386,535 tons of ore that had an average silver content of 4.5 ounces per ton. In addition to copper and silver, the district has produced lesser amounts of gold, lead, and zinc. Since 1957, no metal mines have operated in the Chewelah area. The leading silver-producing mines of the Chewelah area are in table 18.

TABLE 18.—Leading silver mines of the Chewelah area

Mine	Total production (troy ounces)	Average silver content of ore (oz/ton)
United Copper	1,673,072	4
Copper King	39,593	3
Chinto	5,040	33
Eagle (Blue Star)	3,540	10
Jay Gould	2,502	11
Amazon	2,191	3
High Grade (Turk)	1,956	19

Major Mines

United Copper mine

United Copper, which is 3 miles northeast of Chewelah, operated continually from 1906 through 1931, and part time between 1955 and 1957. Since 1957, the mine has been closed. The mine produced a total of 1,673,072 ounces of silver, with a yearly average of 141,288 ounces. In addition to silver, 1,300 ounces of gold, and 9,714,504 pounds of copper was recovered from the mine's total production of 370,554 tons of ore.

The mine is developed over a horizontal distance of 1,500 feet by a 4,220-foot main adit, several hundred feet of sublevels, and a 300-foot shaft. From these workings, which are as much as 1,200 feet beneath the surface, considerable drifting and stoping was undertaken. The vein, which is 3 to 15 feet thick, follows a strong shear zone in argillite. The general strike of the vein is N. 20° E., and its dip is nearly vertical. The milky quartz of the vein contains chalcopyrite, tetrahedrite, and pyrite. The part of the vein that was mined averaged 4 ounces per ton in silver, and 1.5 percent copper; however, one tetrahedrite-rich part of the vein contained 75 to 300 ounces of silver per ton. This high-grade streak was 2 to 12 inches thick, and its horizontal extent was 100 to 200 feet.

Copper King mine

This property adjoins United Copper on the southeast. Between 1904 and 1941, the Copper King produced 13,027 tons of ore that contained 39,593 ounces of silver, 183 ounces of gold, and 446,879 pounds of copper. The mine is developed by several thousand feet of adits and drifts, which contain several stopes. One 200-foot-long stope contained 6,000 tons of ore that averaged 1.5 percent copper. The veins average around 6 feet in thickness; however, bulges in the veins are up to 30 feet thick. The veins consist of quartz and siderite that fill N. 30° E. trending, nearly vertical shear zones in argillite and quartz-mica schist. The sulfide minerals of the veins are chalcopyrite and pyrite, with lesser amounts of tetrahedrite. Tetrahedrite appears to be the main carrier of silver. Several tetrahedrite-rich ore shoots averaged 35 ounces per ton in silver; however, the average silver content for all ore mined was only 3 ounces per ton.

Principal Silver Deposits
of Chewelah District

Amazon

Location: Secs. 29 and 32, T. 33 N.,
   R. 41 E.; part of Chinto property.
Development: 650-foot adit, with 310 feet
   of drifts.
Geology: Quartz veins 2 to 6 feet thick in
   argillite and quartz-mica schist.
   Veins strike N. 23° E. and dip 70°
   NW.
Ore minerals: Chalcopyrite, pyrite, and
   azurite sparsely disseminated in the
   veins.
Production: 1911-1918: 665 tons yielded
   6 ozs. gold, 2,191 ozs. silver, and
   29,540 lbs copper.
References: Fulkerson and Kingston, 1958,
   p. 27; Weaver, 1920, p. 150-151.

Chinto

Location: Near center N½ sec. 32, T. 33 N.,
   R. 41 E.
Development: 1,500-foot adit, and 350-foot
   shaft; total of 7,000 feet of workings
   on three levels.
Geology: Quartz-calcite-siderite veins in
   argillite. Veins strike N. 20° E.,
   have near vertical dips, and are
   up to 6 feet thick.
Ore minerals: Chalcopyrite, tetrahedrite,
   sphalerite, pyrite, arsenopyrite,
   azurite, and malachite occur in lenses
   and as scattered grains.
Production: 1937 and 1939: 151 tons yielded
   9 ozs. gold, 5,040 ozs. silver, and
   15,032 lbs. copper.

References: Fulkerson and Kingston, 1958,
   p. 27; Patty, 1921, p. 131-135.

Copper King

Location: N¼ cor. sec. 32, T. 33 N.,
   R. 41 E.; part of Chinto property.
Development: Several thousand feet of underground workings on two levels. Main
   adit 1,500 feet long.
Geology: Quartz-siderite veins in argillite
   and quartz-mica schist. Veins strike
   N. 30° E.; have near vertical dips;
   and average 6 feet in thickness, with
   lenses up to 30 feet thick.
Ore minerals: Chalcopyrite, tetrahedrite,
   and pyrite in small lenses and as
   disseminated grains.
Production: 1904-1941: 13,027 tons yielded
   183 ozs. gold, 39,593 ozs. silver,
   and 446,879 lbs. copper.
References: Fulkerson and Kingston, 1958,
   p. 27; Patty, 1921, p. 131-135.

Double Eagle

Location: SE¼ sec. 18, T. 31 N., R. 39 E.
Development: 2,200 feet of workings in a
   750-foot adit, a 300-foot adit, and
   a shaft of unknown depth.
Geology: Metallized shear zones in argillite,
   limestone, and quartzite. Boulders of
   solid galena found in overburden. Ore
   averaged 39 ozs. per ton in silver;
   high-grade ore averaged 70 ounces.
Ore minerals: Galena, chalcopyrite, and
   pyrite.
Production: 1916 and 1939: 42 tons yielded
   1,625 ozs. silver, 3,089 lbs. lead,
   and 3,171 lbs. copper.

References: Weaver, 1920, p. 179-180;
Huntting, 1956, p. 240.

Eagle (Blue Star)

Location: Center N½ sec. 5, T. 32 N.,
R. 41 E. and sec. 32, T. 33 N.,
R. 41 E.

Development: Main adit with more than
3,000 feet of drifts and crosscuts,
and a 217-foot shaft.

Geology: Ore shoots in dolomitic limestone
near granodiorite contact. Ore averaged 12 ounces per ton in silver.

Ore minerals: Galena, tetrahedrite, sphalerite, chalcopyrite, cerussite, anglesite,
molybdenite, pyrite, and pyrrhotite.

Production: 1902-1935: 173 tons yielded 2
ozs. gold, 2,166 ozs. silver, 88,693
lbs. lead, and 1,085 lbs. copper.

References: Bancroft, 1914, p. 105-106;
Weaver, 1920, p. 148-149; Huntting,
1956, p. 326.

Edna

Location: SE¼SE¼ sec. 9, T. 31 N., R. 39 E.

Development: 450-foot shaft with 1,000
feet of drifts. Drifts at 110- and 225-
foot levels in shaft.

Geology: 80-foot metallized zone along contact between diorite and argillite.
Ore-grade material averages about
4 feet in width. Vein assays up to
20 ozs. per ton in silver and 7 percent
copper.

Ore minerals: Chalcopyrite, tetrahedrite,
bornite, chalcocite, pyrite, arsenopyrite, and malachite.

Production: 1899, 1904-1906, 1917: $5,000
mainly in copper.

References: Weaver, 1920, p. 181-182;
Huntting, 1956, p. 96.

High Grade

Location: SW¼ sec. 31, T. 33 N., R. 41 E.

Development: 160-foot shaft and a 100-foot
adit with a 50-foot winze.

Geology: Sparsely metallized quartz stringers
up to several inches thick in limestone.

Ore minerals: Chalcopyrite, galena, tetrahedrite, sphalerite, chalcocite,
chrysocolla, malachite, and azurite.

Production: 1917-1920: 98 tons yielded
2 ozs. gold, 1,956 ozs. silver, and
14,914 lbs. copper.

References: Fulkerson and Kingston, 1958,
p. 28, Weaver, 1920, p. 144.

Jay Dee

Location: SW¼ sec. 31, T. 33 N., R. 41 E.

Development: 160-foot shaft, and a 100-foot
adit with a 50-foot winze.

Geology: Narrow quartz stringers up to several inches thick in limestone.

Ore minerals: Galena, tetrahedrite, sphalerite, malachite, azurite, chalcocite,
and chrysocolla.

Production: 1917, 1919, 1920: 98 tons of
ore yielded 2 ozs. gold, 1,956 ozs.
silver, and 14,914 lbs. copper.

References: Weaver, 1920, p. 144-145;
Huntting, 1956, p. 328.

Jay Gould

Location: Center S½ sec. 8, T. 32 N.,
R. 41 E.

Development: 200-foot inclined shaft with
two levels containing several hundred
feet of drifts.

Geology: Quartz veins in argillite near granite contact. Ore averaged 11 ounces
per ton in silver.

Ore minerals: Galena, sphalerite, chalcopyrite, tetrahedrite, bornite, and pyrite.

Production: 1902-1904, 1907, 1939: 226
tons yielded 2,502 ozs. silver, 42,742
lbs. lead, and 82 lbs. copper.

References: Bancroft, 1914, p. 106-107;
Weaver, 1920, p. 151-152.

Krug (Hartford)

Location: NE¼SE¼ sec. 26, T. 33 N.,
R. 39 E.

Development: 340-foot adit, 35-foot shaft,
and several open cuts.

Geology: Several quartz veins in limestone
and diabase. One 2- to 3-foot thick
vein is fairly well metallized, and
assays up to 94.2 ozs. of silver per
ton.

Ore minerals: Chalcopyrite, tetrahedrite,
galena, sphalerite, pyrite, malachite,
azurite, and cuprite.

Production: 1916-1917: 78 tons yielded
316 ozs. silver and 52 lbs. copper.

References: Weaver, 1920, p. 167-169.

Mullen

Location: W½NW¼ sec. 16, T. 32 N.,
R. 41 E.

Development: 17-foot shaft.

Geology: 3- to 4-foot-wide quartz vein
along granodiorite-quartzite contact.
High-grade ore contains 34 percent
lead, 8 percent zinc, 21.6 ozs. silver,
and 0.56 ozs. gold.

Ore minerals: Galena, sphalerite, tetrahed-

rite, zinkenite, chalcopyrite, and
pyrite.

References: Huntting, 1956, p. 245;
Purdy, 1951, p. 142-143.

Nevada

Location: SW¼ sec. 23, T. 32 N., R. 39 E.

Development: 30-foot shaft.

Geology: Ore minerals form pods and pockets
in limestone.

Ore minerals: Galena, sphalerite, tetrahedrite, chalcopyrite, and pyrite.

Production: 1937: 5 tons of ore yielded 58
ozs. silver, 1,390 lbs. lead, and 38
lbs. copper.

References: Huntting, 1956, p. 245;
Fulkerson and Kingston, 1958, p. 28.

U.S. Copper Gold

Location: SE¼SE¼ sec. 8, T. 32 N., R. 41 E.

Development: 340-foot adit.

Geology: 4-foot-thick quartz vein in quartzite, limestone, and argillite.

Ore minerals: Chalcopyrite and pyrite.

Production: 1916-1917: 20 tons of ore
yielded 35 ozs. silver and 1,104 lbs.
copper.

References: Weaver, 1920, p. 161; Huntting,
1956, p. 333.

United Copper

Location: E½NW¼ sec. 32, and NE¼ sec. 31,
T. 33 N., R. 41 E.

Development: Main adit 4,200 feet long and
a 300-foot shaft with considerable
drifting and stoping on several levels.

Geology: Quartz-calcite-siderite vein 5 to
20 feet thick in argillite. Vein

strikes N. 20° E., and is nearly vertical. High-grade ore contained 75
to 300 ounces of silver per ton.

Ore minerals: Chalcopyrite, argentian tetrahedrite, pyrite, arsenopyrite, and
malachite.

Production: 1906-1931, 1955-1957: 370,554
tons of ore yielded 1,673,072 ozs.
silver, 1,300 ozs. gold, and 9,714,504
lbs. copper.

References: Patty, 1923, p. 123-131;
Weaver, 1920, p. 139-141.

Wells Fargo

Location: NW¼NE¼ sec. 36, T. 31 N.,
R. 38 E.

Development: Main adit 1,500 feet long with
short drifts and raise. Upper adit contains 200 feet of workings.

Geology: 3- to 5-foot-thick quartz-barite
veins in argillite and dolomite.
Zinkenite contains up to 25 ozs. per
ton in silver, 24 percent antimony,
18 percent lead, and 18 percent zinc,
and occurs as solid masses up to 18
inches thick.

Ore minerals: Zinkenite, sphalerite, stibnite,
jamesonite, and pyrite.

Production: 1916: two carloads of barite.
1960: 50 tons barite.

References: Purdy, 1951, p. 146-148; Moen,
1964, p. 53-55.

DEER TRAIL DISTRICT

Location

The Deer Trail district is in southwestern
Stevens County, in the general vicinity of Fruitland
(fig. 10). The principal lead-silver deposits are at
the head of Cedar Canyon, and near the headwaters
of O-Ra-Pak-En Creek. Elevations in the district
range from 3,600 to 4,650 feet. Deer Trail is also
known as the Cedar Canyon district.

Geology and Mineralization

The silver-bearing deposits of the Deer Trail
district occur mainly as quartz veins in a steeply
dipping, northeast-trending belt of dolomite, slate,
and phyllite, which is a part of the Deer Trail Group
(Precambrian). The rocks have been intruded by

quartz monzonite of the Loon Lake batholith (Cretaceous). Most veins occupy moderately to steeply
dipping northeast-trending shear zones that parallel
several major high-angle faults. The veins consist
mainly of white quartz that is accompanied by varying amounts of calcite, siderite, and barite. The
veins range from less than a foot to as much as 12 feet
in thickness, with an average of 2 to 3 feet. Metallized parts of the veins contain pods and lenses of
ore minerals that are as much as 6 feet thick. However, mining in the past indicates that the average
ore shoot is only around 18 inches thick. The silver
minerals of the veins consist of argentite, cerargyrite,
stephanite, argentiferous galena, and native silver.
These minerals are accompanied by sphalerite, chalcopyrite, tetrahedrite, azurite, malachite, pyrite,
and limonite. Some silver-bearing veins have been
mined for several hundred feet along their strikes,

R 37E

Mines and Prospects

1. Orozada
2. Queen-Seal
3. Deer Trail
4. Aichan Bee
5. Brooks
6. Orchid
7. Saturday Night – Sunday Morning
8. Silver Star

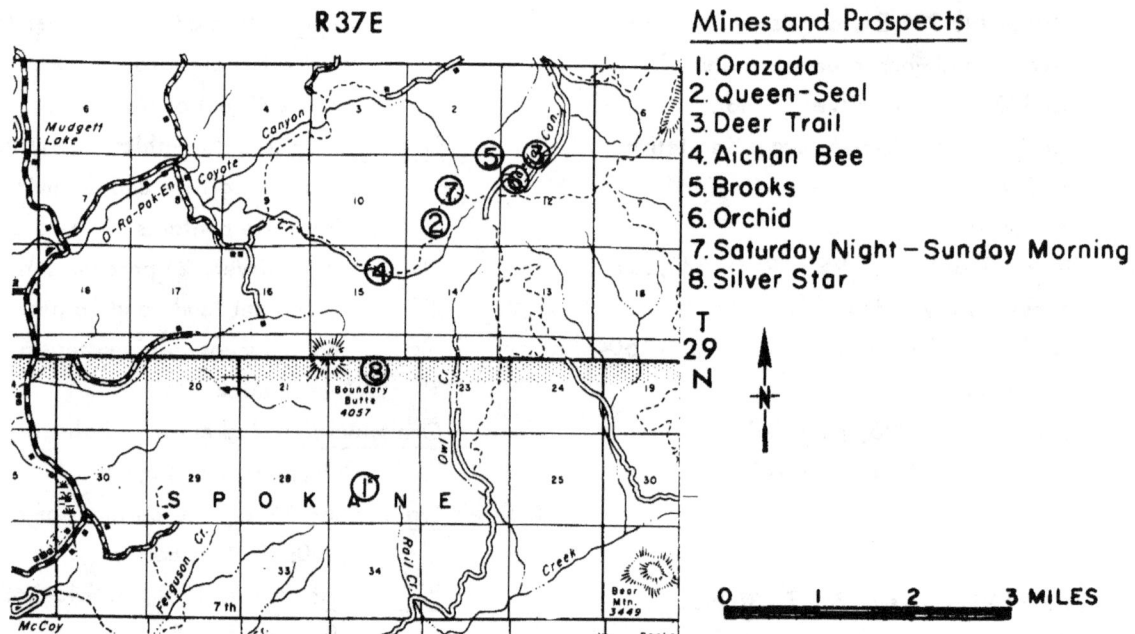

FIGURE 10.—Index map of the Deer Trail district.

and for a maximum depth of about 400 feet beneath their outcrops; however, nothing suggests that mineralization terminates at this depth.

The average silver content of ore from the district was 9.82 ounces per ton; however, individual shipments of ore were much richer. At the Legal Tender mine, 122,211 tons of ore averaged 216.30 ounces per ton in silver, while 881 tons of ore from the Queen-Seal mine averaged 240.3 ounces per ton. Properties that have high lead contents generally have a high silver content. The silver content of lead ore from mines in Cedar Canyon ranged from 13.5 to 215 ounces per ton, with an average of 150 ounces per ton. At properties where copper minerals predominate, the silver content of the ore is lower. At the Turk mine, where the chief copper mineral is chalcopyrite, the average silver content of the ore is only 0.32 ounce per ton. At the Queen-Seal, where tetrahedrite is the main copper mineral, the copper ore averaged about 16 ounces per ton in silver. At the Prov-

idence and Queen-Seal, parts of the veins contained mainly silver minerals and little if any ore minerals of copper, lead, and zinc. The silver content of these veins was as much as 200 ounces per ton.

Production

Although not the most productive silver area of Stevens County, some of the richest silver ore mined in the state came from the Deer Trail district. As early as 1894, mines in Cedar Canyon shipped ore that averaged 200 ounces per ton in silver and 8 percent lead; by 1900 the production of lead and silver reached $500,000. The district's most productive mines were the Legal Tender, Providence, Queen, and Silver Seal. From 1902 through 1967, mines of the area produced 21,606 tons of ore that contained 89 ounces of gold, 190,602 ounces of silver, 292,780 pounds of copper, 1,874,066 pounds of lead, and

558,800 pounds of zinc. Sixteen mines in the district have produced silver; however, only three mines have been major silver producers. The leading silver mines of the Deer Trail district are the Legal Tender, Providence, and Queen-Seal; currently (1975) all mines are idle, except for limited development work at the Deer Trail mine.

TABLE 19. — Leading silver producers of the Deer Trail district (1894-1941)

Mine	Total production (troy ounces)	Average silver content of ore (oz./ton)
Queen-Seal	285,759	97.53
Legal Tender	122,211	216.30
Providence	87,442	143.80

Major Mines

Queen-Seal mine

This property, which is near the headwaters of O-Ra-Pak-En Creek, consists of the Silver Queen and Silver Seal mining claims from which rich silver ore was first shipped in 1900. In 1902, ore shipped from the Seal averaged 240 ounces per ton in silver, and ore from the Queen averaged 200 ounces. From 1900 through 1941, the total combined production of the Queen-Seal was 2,930 tons of ore that averaged 97.53 ounces per ton in silver. In addition to 285,759 ounces of silver, the mine produced 14,148 pounds of lead, and 821 pounds of copper. Almost all the ore was shipped crude to smelters; however, in 1937 and 1938, a 50-ton flotation mill was operated at the mine by Queen-Seal Mining Co. Small-scale mining was attempted in the 1960's, but less than 100 tons of ore was shipped.

The mine contains several thousand feet of workings that are inaccessible because of caved adits and shafts. The main adit intersects the vein about 400 feet from its portal, and two shafts were sunk on the vein for depths of 300 and 200 feet beneath the surface; considerable stoping was undertaken on the vein, which has been drifted upon for over 1,600 feet. The silver-bearing quartz vein is 1 to 3 feet thick, strikes N. 25°-35° E., and dips 75° NW. Ore shoots, which pitch to the northeast, contain argentite, native silver, tetrahedrite and cerargyrite that is accompanied by minor pyrite, galena, and sphalerite. Host rocks for the vein consist of massive white dolomite and artillite, both of the Deer Trail Group (Precambrian). These rocks are no doubt underlain by Cretaceous age granodiorite and quartz monzonite, as the granitic rocks crop out less than 1 mile both east and west of the Queen-Seal mine.

Deer Trail mine

The Deer Trail mine at the head of Cedar Canyon is reported to have produced more than $3 million from 1894 to 1947, with the most productive years being 1894 to 1911. Some ore contained as much as 2,000 ounces per ton in silver, and in the late 1890's and early 1910's only ore that ran 80 ounces or more per ton in silver was shipped; the shipments averaged 300 ounces per ton in silver and 8 percent lead. Prior to 1900, most production was from the Deer Trail claims, whereas, from 1900 to 1911, production was mainly from the Elephant, Legal Tender, and Providence claims of the Deer Trail group. From 1902 through 1911, 1,173 tons of ore shipped from the Legal Tender and Providence averaged 265 ounces per ton in silver, and 11 percent lead. The mine has not been in production since 1947, and the portals of most adits are caved. Currently (1975), the property is under development by Deer Trail Mines Co.

The mine is developed over a vertical distance of 500 feet by 11 adits that contain over 4,000 feet of drifts and raises. In several parts of the mine, the vein has been extensively stoped. Near-surface parts of the vein contained abundant limonite rich in cerargyrite and native silver. Below the oxidized zone the chief ore minerals were argentiferous galena, argentite, and sphalerite. The ore minerals occur in quartz-calcite veins that follow shear zones in argillite and dolomite of the Deer Trail Group (Precambrian). The veins have a general N. 35°-50° E. strike and dip 35° to 60° SE.; they average around 30 inches in thickness, and contain tabular and lenticular ore shoots that are up to $2\frac{1}{2}$ feet thick. Generally, the thinnest ore shoots were richest in silver. In the vicinity of the mine, rocks of the Deer Trail Group are underlain by Cretaceous granodiorite and quartz monzonite. These granitic rocks are probably not much more than 1,500 feet beneath the mine workings, and are probably the source rocks for the silver deposits of the area.

Because all adits are caved at their portals, it is not possible to determine if ore remains in the mine. According to Max Slate, owner of the mine, the old mine dumps have been sampled, and average around 15 ounces per ton in silver, while parts of some veins still contain silver ore.

Principal Silver Deposits of Deer Trail District

Aichan Bee

Location: Center NE$\frac{1}{4}$ sec. 15, T. 29 N., R. 37 E.

Development: 600-foot adit with a 150-foot drift.

Geology: Ore minerals disseminated along shear zones in argillite, quartzite, and limestone.

Ore minerals: Sphalerite, galena, chalcopyrite, and pyrite. Ore assays up to 12 percent lead, 18 percent zinc, and 12 ozs. per ton in silver.

Production: Minor amounts in 1943.

Reference: Huntting, 1956, p. 323.

Brooks

Location: NE$\frac{1}{4}$ cor. sec. 11, T. 29 N., R. 37 E.

Development: 50-foot shaft and 2 adits.

Geology: 5- to 7-foot quartz veins in argillite and limestone.

Ore minerals: Stephanite, galena, sphalerite, tetrahedrite, and pyrite.

Production: Shipped ore prior to 1901.

Reference: Huntting, 1956, p. 324.

Cleveland

Location: NE$\frac{1}{4}$ sec. 9, T. 30 N., R. 38 E.

Development: Lower adit with 1,400 feet of crosscuts, drifts, and raises; intermediate adit 750 feet long; upper adit with 1,200 feet of drifts, crosscuts, and raises. Several other adits, shafts, and open cuts.

Geology: Lenses of ore in brecciated dolomitic limestone. Ore also occurs in thin veins and in chimneys. Ore averaged 3.73 ozs. per ton in silver, 5 percent lead, and 1 percent zinc. High-grade ore contained up to 132 ozs. per ton in silver.

Ore minerals: Galena, stibnite, boulangerite, sphalerite, tetrahedrite, chalcopyrite, cerussite, anglesite, bindheimite, pyrite, arsenopyrite, and malachite.

Production: 1903-1948: 26,459 tons yielded
89 ozs. gold, 98,745 ozs. silver,
2,700,000 lbs. lead, 551,170 lbs.
zinc, 7,396 lbs. copper. 1968: 56 tons.

References: Purdy, 1951, p. 137-141;
Jenkins, 1924, p. 127-132; Gage,
1941, p. 91-96.

Deer Trail

Location: SW$\frac{1}{4}$ sec. 1 and NW$\frac{1}{4}$ sec. 12,
T. 29 N., R. 37 E.

Development: 11 adits containing over 4,000
feet of drifts, crosscuts, and raises.
Vein extensively stoped in several
adits.

Geology: 1- to 6-foot-thick quartz vein in
argillite and dolomite containing
tabular and lenticular ore shoots up
to 2$\frac{1}{2}$ feet thick. High-grade shoots
contained up to 2,000 ozs. per ton in
silver. Ore averaged around 280 ozs.
per ton in silver, and 10 percent lead.

Ore minerals: Cerargyrite, native silver,
argentite, galena, pyrargyrite,
sphalerite, cerussite, tetrahedrite,
and pyrite.

Production: 1894-1902: substantial production of unknown value. 1902-1947:
1,855 tons of ore yielded 2 ozs. gold,
304,557 ozs. silver, 323,417 lbs.
lead, and 293 lbs. copper.

References: Huntting, 1956, p. 325; Bancroft,
1914, p. 117-118; Jenkins, 1924,
p. 134-136; Poole, 1936.

Indian Trail

Location: NE$\frac{1}{4}$ sec. 31, T. 29 N., R. 36 E.

Development: 400-foot crosscut with drift
and 150-foot raise to surface.

Geology: 2-foot-thick quartz vein in mon-
zonitic and metasedimentary rocks.
Vein carries 15 to 20 ozs. of silver,
6 to 7 percent lead, and 4 to 5 percent zinc.

Ore minerals: Galena, sphalerite, and pyrite.

Production: Small shipment in 1969 assayed
21.6 ozs. silver, 0.05 oz. gold, 7.4
percent lead, and 4.7 percent zinc.

Reference: Washington Division of Geology
and Earth Resources files.

Orchid

Location: Sec. 12, T. 29 N., R. 37 E.

Development: 1,750 feet of workings in shaft
and several adits.

Geology: Quartz vein in argillite and limestone.

Ore minerals: Argentite and cerargyrite.

Production: 3 tons in 1900.

Reference: Huntting, 1956, p. 330.

Orazada

Location: SE$\frac{1}{4}$NW$\frac{1}{4}$ sec. 27, T. 39 N., R. 37 E.,
on Spokane Indian Reservation.

Development: 2,400-foot adit.

Geology: Narrow metallized shear zones in
argillite and limestone. Ore averaged
13.5 ozs. per ton in silver.

Ore minerals: Galena, sphalerite, jamesonite,
pyrite, and arsenopyrite.

Production: 1927 and 1941: 35 tons of ore
yielded 468 ozs. silver, 5 ozs. gold,
7,374 lbs. lead, and 77 lbs. copper.

References: Gage, 1941, p. 110-111;
Huntting, 1956, p. 330.

Queen-Seal

Location: SW¼ sec. 11, T. 29 N., R. 37 E.

Development: 300- and 200-foot shafts, and a 1,370-foot crosscut adit; several stopes, and 1,500 feet of drifts on the vein.

Geology: 1- to 3-foot-thick quartz vein in dolomite and argillite. Vein strikes N. 25°-35° E., dips 75° NW.

Ore minerals: Argentite, cerargyrite, native silver, galena, sphalerite, tetrahedrite, and pyrite. High-grade ore contained up to 240 ozs. per ton in silver; much ore averaged 97.53 ozs. per ton in silver and 10 percent lead.

Production: 1900-1941: 2,930 tons of ore yielded 285,759 ozs. silver, 14,148 lbs. lead, and 821 lbs. copper.

References: Weaver, 1920, 193-195; Huntting, 1956, p. 331.

Saturday Night-Sunday Morning

Location: Sec. 11, T. 29 N., R. 37 E.

Development: Caved adits and shafts with 800 feet of workings.

Geology: Metallized zone, 18 inches thick in argillite and limestone.

Ore minerals: Tetrahedrite, galena, and sphalerite.

Production: Two tons of ore shipped prior to 1897 contained 71 ozs. of silver.

Reference: Huntting, 1956, p. 247.

Silver Star

Location: NE¼SW¼ sec. 22, T. 29 N., R. 37 E.

Development: Several open cuts and a caved shaft.

Geology: Quartz vein up to 5 feet thick in dolomite. Vein averages 8 ozs. per ton in silver.

Ore minerals: Galena and tetrahedrite.

Production: None.

Reference: Washington Division of Geology and Earth Resources files.

BOSSBURG DISTRICT

Location

The Bossburg district of northwestern Stevens County lies between the town of Bossburg on the west, and the headwaters of Clugston Creek on the east. The area includes parts of T. 37 N., R. 38 and 39 E. (fig. 11). The area is mountainous, and fairly well accessible. Elevations range from 1,380 feet at Bossburg to a maximum of 4,655 feet on Uncle Sam Mountain in the eastern part of the area. Lake Roosevelt is at the western edge of the area; Bruce Creek is the principal stream in the central part of the area, and Clugston Creek is the principal stream in the eastern part.

Geology and Mineralization

Silver in the Bossburg district is associated mainly with lead-zinc deposits. The host rocks for the deposits are limestone, dolomite, argillite, quartzite, and schist, of late Paleozoic and Triassic age. Mesozoic and Tertiary granitic rocks underlie parts of the area, and appear to be the source of the metal deposits. In order of decreasing abundance the ore minerals are galena, sphalerite, cerussite, smithsonite, chalcopyrite, tetrahedrite, and geochronite. Minor calamine, wulfenite, anglesite, azurite, and malachite occur at several deposits, while pyrite and

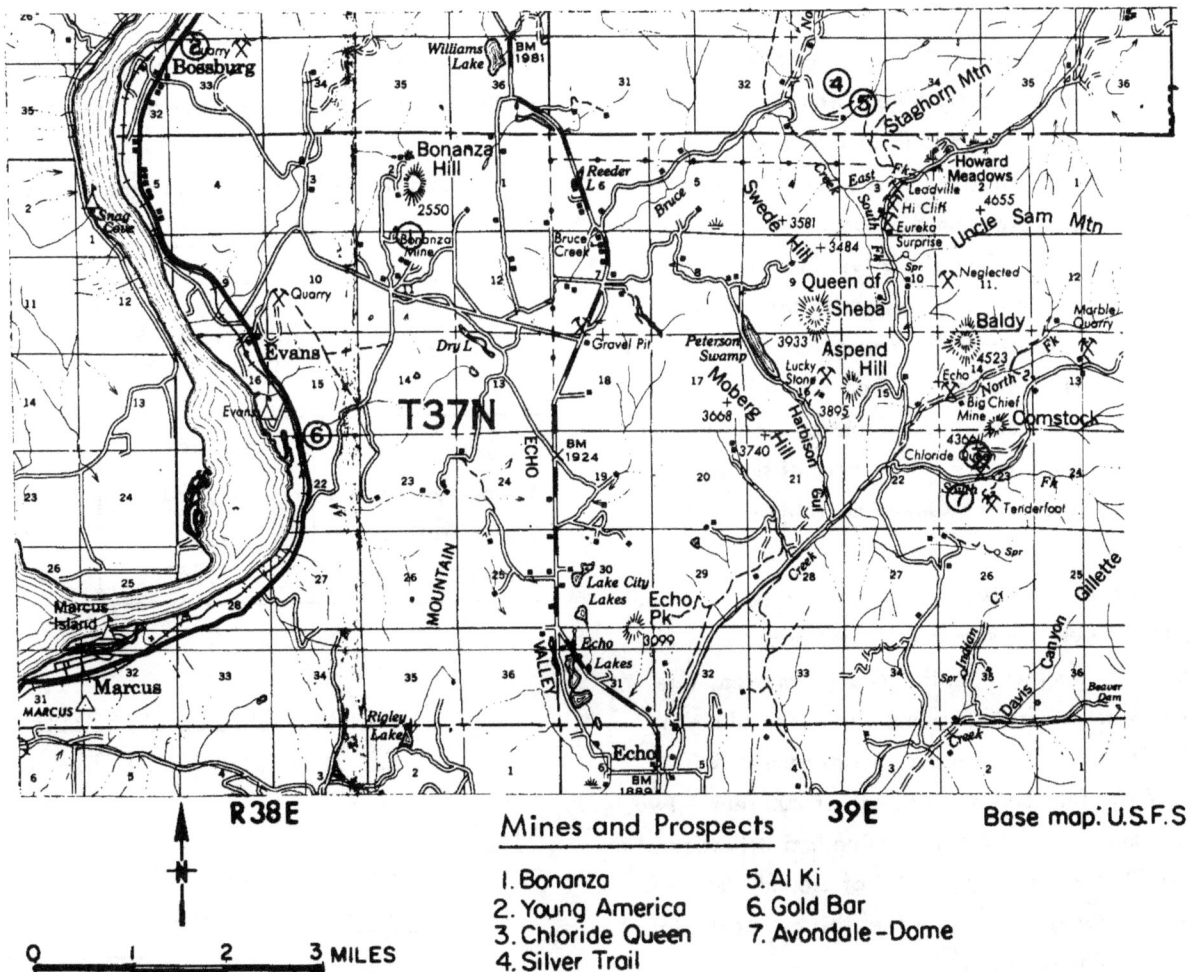

Mines and Prospects

1. Bonanza   5. Al Ki
2. Young America   6. Gold Bar
3. Chloride Queen   7. Avondale-Dome
4. Silver Trail

Base map: U.S.F.S

FIGURE 11.—Index map of the Bossburg district.

limonite are present at most deposits. Silver minerals have not been reported in any ores of the Bossburg area; the silver occurs mainly in argentiferous galena and argentian tetrahedrite.

The ore minerals occur mainly in nodules, pods, and stringers in quartz-calcite veins. The veins, which average 4 to 5 feet in thickness, occur along shear zones, and are not persistent for great distances. Other pods and nodules of ore minerals are randomly scattered in fractured and sheared limestone, and form disseminated deposits, some of which are chimneylike in shape. At the Young America mine, disseminated grains of sphalerite and galena occur along bedding planes in limestone, whereas at the Bonanza mine galena occurs as replacement lenses along schistosity planes in schist. At the Uncle Sam mine, galena and sphalerite occur as irregular shaped masses in a chimney of highly brecciated dolomite.

At most properties in the Bossburg area, oxidation at the surface is only slight. Some sphalerite has been altered to smithsonite, and cerussite is the common alteration product of galena.

The average silver content of ore from the Bossburg area was only 2.63 ounces per ton. This figure is influenced greatly by ore from the Bonanza mine, which produced the bulk of lead and silver from the Bossburg area. The average silver content of 101,994 tons of ore from the Bonanza was 2.34 ounces per ton. Ore from the Silver Trail mine averaged 7 ounces per ton in silver, while the average silver content of ore from the Young America mine was 5.22 ounces. Hand-sorted ore from several other mines was much richer. Ninety tons of ore from the Minorca mine contained 9,259 ounces of silver, and 35 tons from the Chloride Queen contained 1,569 ounces.

Most ore shoots in the Bossburg area have been small; the largest ore shoots were at the Bonanza and Young America mines. In the Young America, one ore shoot was 300 feet long and was mined along its dip for 200 feet. Another ore shoot had a strike length of 100 feet and a dip length of 200 feet. Two main ore shoots at the Bonanza mine had strike lengths of up to 200 feet and dip lengths of around 600 feet. The shoots averaged around 4 feet thick, but in places were up to 27 feet thick.

Production

Production of silver from the Bossburg district between 1902 and 1956 totaled 307,288 ounces. Prior to 1902, and as early as 1885, the Young America, Bonanza, Silver Trail, and Chloride Queen mines had a combined silver production of around 100,000 ounces. Peak production years for the district were 1947 through 1952; silver production averaged 33,536 ounces yearly, with most of the silver coming from the Bonanza mine. Between 1903 and 1954, mines of the district produced 116,961 tons of ore that contained 102 ounces of gold, 307,288 ounces of silver, 145,765

pounds of copper, 24,785,332 pounds of lead, and 971,514 pounds of zinc. The leading silver producers of the Bossburg district are shown in table 20.

TABLE 20.—Leading silver producers of the Bossburg district

Mine	Total production (troy ounces)	Average silver content of ore (oz./ton)
Bonanza	238,485	2.34
Young America ..	69,893	5.22
Silver Trail	2,916	7.00
Chloride Queen ..	1,569	44.80

Principal Silver Deposits of Bossburg District

Al Ki

Location: Sec. 33, T. 38 N., R. 39 E.

Development: Unknown.

Geology: 3- to 4-foot quartz veins in granite. Select samples from vein contained 17 to 56 ozs. of silver, 2.5 ozs. gold, and 50 to 64 percent lead.

Ore minerals: Galena and pyrite.

Production: Unknown.

References: Huntting, 1956, p. 236.

Avondale-Dome

Location: SW$\frac{1}{4}$ sec. 2, T. 37 N., R. 39 E.

Development: 800 feet of workings in 108- and 125-foot adits.

Geology: Lenses and nodules of ore minerals in a 4-foot shear zone in limestone. Select samples assayed 60 percent lead and 15 ozs. silver.

Ore minerals: Galena with minor cerussite, calamine, anglesite, wulfenite, limonite, and pyrite.

Production: Unknown.

References: Huntting, 1956, p. 237.

Bonanza

Location: Secs. 2 and 11, T. 37 N., R. 38 E.

Development: 750-foot incline shaft with seven working levels; more than 15,000 feet of underground workings; metallized shear zone extensively stoped.

Geology: Ore shoots up to 27 feet thick along shear zone in chlorite and graphite schist. Ore shoots lenticular, and averaged 3 to 4 feet in thickness. Average silver content 2.34 ozs. per ton.

Ore minerals: Galena, sphalerite and pyrite.

Production: 1907-1952: 101,994 tons yielded 12 ozs. gold, 238,485 ozs. silver, 24,880,359 lbs. lead, 230,687 lbs. zinc and 21,913 lbs. copper. Prior to 1907 several thousand tons of unknown value.

References: Weaver, 1920, p. 174-176; Jenkins, 1924, p. 118-120; Bancroft, 1914, p. 63-64; Huntting, 1956, p. 238; Patty, 1921, p. 117-119.

Chloride Queen

Location: E$\frac{1}{2}$ sec. 23 and W$\frac{1}{2}$ sec. 24, T. 37 N., R. 39 E.

Development: 350-foot adit with drifts, shafts and stopes; 1,500-foot adit with short crosscuts.

Geology: 5-foot quartz vein along quartzite-dolomite contact. Also replacement deposits in brecciated limestone. Several ore shoots contained up to 34.3 percent lead, and 14 ozs. of silver per ton. One ore shoot contained 136 ozs. per ton in silver.

Ore minerals: Galena, cerussite, sphalerite, chalcopyrite, limonite, and pyrite.

Production: 1900, 1914, 1924, 1929. Eleven tons of ore mined in 1929 yielded 1 oz. gold, 1,492 ozs. silver, 10 lbs. copper and 160 lbs. lead.

References: Jenkins, 1924, p. 113-115; Gage, 1941, p. 43-45; Weaver, 1920, p. 235-237.

Gold Bar

Location: Secs. 15 and 22, T. 37 N., R. 38 E.

Development: 1,000 feet of underground workings in 4 adits and 2 shafts.

Geology: Narrow quartz veins fill fractures in quartzite and argillite. Veins assay from 1.2 to 62 ozs. silver, trace to 1.3 percent lead, trace to 1.7 percent zinc, trace to 0.3 percent copper, and 1.02 to 2.70 percent arsenic.

Ore minerals: Galena, sphalerite, tetrahedrite, pyrite, and arsenopyrite.

Production: Unknown.

Reference: Huntting, 1956, p. 169.

Silver Trail

Location: N$\frac{1}{2}$ sec. 33, T. 38 N., R. 39 E.

Development: 850- and 320-foot adits, and two shallow shafts; several open cuts.

Geology: 1½- to 8-foot-thick quartz veins
in argillite and quartzite. High-grade
ore contained 20 percent lead, and
129 ozs. per ton in silver. Average
ore contained 7 ozs. per ton in silver,
and 1.5 percent lead.

Ore minerals: Galena, sphalerite and chalcopyrite.

Production: About 6,000 tons of ore from
1891 to 1952; 908 tons produced
from 1912 to 1952 contained 4,876
ozs. silver, 26,333 lbs. lead, and
27,701 lbs. zinc.

References: Weaver, 1920, p. 246-248;
Bethune, 1892, p. 127-128;
Huntting, 1956, p. 248.

Young America

Location: NE¼NW¼ sec. 33, T. 38 N.,
R. 38 E.

Development: Six adits with considerable
stoping.

Geology: Disseminated ore minerals and
stringers and lenses of ore minerals
in dolomite. Area of disseminated
ore minerals 30 to 150 feet wide and
900 feet long. Ore mined to date
averaged 5.2 ozs. per ton in silver,
3.5 percent lead, and 2.9 percent
zinc.

Ore minerals: Galena, sphalerite, smithsonite, cerussite, geocronite, and
pyrite.

Production: 1905-1953: 13,409 tons yielded
50 ozs. gold, 70,857 ozs. silver,
5,138 lbs. copper, 949,719 lbs.
lead, and 771, 629 lbs. zinc.
Produced between 1889 and 1905,
but value of ore unknown.

References: Hundhausen, 1949, 13 p.;
Patty, 1921, p. 119-121; Purdy,
1951, p. 148-153.

SUMMIT DISTRICT

Location

The Summit district of west-central Stevens
County lies immediately east of Roosevelt Lake; Kettle
Falls district is to the north, and Deer Trail district
is to the south (fig. 8). The Summit district is classed
as a lead-silver district; however, production from
the district has been minor.

Geology and Mineralization

The ore deposits of the Summit district occur
mainly as metallized quartz-calcite fissure veins in
Paleozoic argillite and quartzite, which have been
intruded by Mesozoic diorite, porphyritic granodiorite, and trachite. The veins, which have northwest
to northeasterly strikes and nearly vertical dips, are
as much as 12 feet thick; the average thickness is 3
to 4 feet. The veins commonly pinch and swell, and
some veins are offset several feet by post-mineraliza-
tion faulting. Common ore minerals, sparsely disseminated in most veins, are argentiferous galena,
sphalerite, and chalcopyrite. The less common ore
minerals are tetrahedrite, scheelite, wolframite,
argentite, molybdenite, cerussite, and anglesite.
The ore minerals are almost always accompanied by
pyrite, and occasionally by arsenopyrite and marcasite. Although the ore minerals are only sparsely disseminated in the veins, parts of some veins contain
pods and lenses of ore minerals that are up to several
feet thick.

Production

Almost the total production of silver, gold, copper, and lead of the Summit district has come from the Daisy mine. Prior to 1890, the mine produced 2,000 tons of ore of undisclosed value. Sporadic mining operations from 1916 through 1935 yielded 1,484 tons, averaging 9.43 ounces of silver per ton and 1 percent lead. Around 1890, high-grade lead-silver ore was shipped from the Silver Summit mine to Denver smelters. The ore averaged 50 ounces per ton in silver and 45 percent lead. In 1935, 2 tons of ore from the Silver Summit yielded 26 ounces per ton in silver and 29 percent lead. There has been no production from the district since 1935; however, exploration work was undertaken at the Daisy mine in 1970, and at the Silver Summit mine in 1968.

Principal Silver Deposits of Summit District

Daisy-Tempest

Location: Secs. 6 and 7, T. 33 N., R. 38 E.

Development: 11 adits with more than 4,000 feet of underground workings over a vertical distance of 600 feet.

Geology: Quartz-calcite veins up to 16 feet thick along shear zones in argillite and limestone. Main vein strikes N. 52° W., dips 70° SW. and contained ore shoots as much as 2½ feet in thickness, which averaged around 20 ozs. per ton in silver.

Ore minerals: Galena, argentite, chalcopyrite, sphalerite, cerargyrite, pyrargyrite, azurite, malachite, pyrite, and arsenopyrite.

Production: Main production prior to 1900; 1,482 tons produced from 1916 through 1935 yielded 10 ozs. gold, 13,978 ozs. silver, 4,527 lbs. copper, and 28,967 lbs. lead.

References: Patty, 1921, p. 136-143; Jenkins, 1924, p. 124-126.

Silver Summit

Location: Center sec. 33, T. 33 N., R. 37 E.

Development: 3 shafts, 1 adit, and several open cuts. One incline shaft 300 feet deep with 340 feet of drifts on the vein.

Geology: 2½- to 5-foot-thick quartz vein in trachite. Vein strikes N. 30° E., dips 35° NW. Ore minerals occur in scattered pods and pockets. Garnet and epidot near contact of trachite and limestone.

Ore minerals: Galena, pyrite, and anglesite; some scattered grains of scheelite and powellite.

Production: Main production around 1890 but probably not much more than 100 tons. Two tons in 1935 yielded 52 ozs. silver, 1,178 lbs. lead, and 14 lbs. copper.

References: Bethune, 1891, p. 76-77; Huntting, 1956, p. 248.

COLVILLE DISTRICT

Location

The Colville district of central Stevens County centers about Old Dominion Mountain, the summit of which is 7 miles east of Colville (fig. 8). The only significant silver mine in the area is the Old Dominion, which was the first major silver-lead discovery in Washington, and a producer of some of the richest silver ore mined in the state. Old Dominion mine is on the southwest slope of the mountain where elevations are from 2,925 to 3,375 feet above sea level.

Geology and Mineralogy

Ore deposits of the Colville area occur as metallized quartz fissure veins, and as replacement deposits along shear and breccia zones. Host rocks for the ore deposits consist mainly of Cambrian limestone, quartzite, and schist; also, several deposits occur in Precambrian greenstone and Mesozoic granodiorite. The quartz veins range from several inches to 6 feet in thickness, whereas, the metallized shear and breccia zones are up to 10 feet thick. Most deposits are steeply dipping, but at the Old Dominion several ore bodies dip as little as 10 to 20 degrees, and are irregular in size and shape. The principal gangue mineral is massive white quartz, which is often accompanied by calcite, dolomite, and siderite. The common ore minerals are argentiferous galena and sphalerite that are almost always accompanied by pyrite, and occasionally by arsenopyrite. Some deposits contain tetrahedrite and scheelite; near-surface ore from the Old Dominion mine contained cerussite, anglesite, stephanite, and native silver. The ore shoots in most deposits are small; however, several ore shoots in the Old Dominion mine were 6 to 8 feet thick and had strike and dip lengths of several hundred feet. Several ore bodies in the mine appear to have been deposited at the intersection of shear zones.

The average silver content of most veins did not exceed 10 ounces per ton. However, at the Old Dominion, near-surface ore mined in the late 1890's and early 1900's contained as much as 478 ounces per ton in silver, while ore mined from 1906 to 1931 in the lower workings of the mine averaged 87.78 ounces per ton.

Production

The Colville district is the second most productive silver district in Stevens County. From 1902 through 1954, mines of the district produced 6,752 tons of ore that contained 370 ounces of gold, 362,467 ounces of silver, 5,371 pounds of copper, 960,761 pounds of lead, and 405,419 pounds of zinc. Sixty percent of the total metal production came from the Old Dominion, and the value of silver produced at the mine exceeded the value of any other metal. Other major producers were the Longshot and Shoemaker mines, both of which produced several thousand tons of lead-zinc ore.

Major Mines

Old Dominion mine

This property, which is 6 miles east of Colville, was one of the first major silver mines in Washington and operated sporadically from 1885 until 1953. From 1885 to 1895, about $500,000 in high-grade lead-silver ore was mined from near-surface ore shoots that averaged 400 ounces per ton in silver, and 33 percent lead. Ore mined from 1902 through 1953, averaged 83.78 ounces per ton in silver and 9.1 percent lead. In 1952 and 1953, the silver content of the ore had dropped to 13.5 ounces per ton, and the lead content was only 2.15 percent. From 1885 through 1953, the mine produced 6,588 tons of ore that contained 323 ounces in gold, 942,517 ounces of silver, 2,244,391 pounds of lead, 148,563 pounds of zinc, and 4,132 pounds of copper. The mine has been idle since 1953; some underground workings are caved, and nothing remains of the concentrating mill.

The property is developed by about 8 miles of underground workings on at least 11 levels that extend over a vertical distance of about 600 feet. No. 1 adit is the lowest adit, and its portal is at 2,975 feet elevation. The adit is over 5,500 feet long, and ore has been mined from several thousand feet of sublevels, most of which lie up to 200 feet beneath the No. 1 adit level. Extensive longhole exploratory drilling has been undertaken on all levels in search of ore bodies, and the ore bodies appear to have bottomed at about 200 feet beneath the No. 1 adit level. No. 2 adit is at an elevation of 3,130 feet, and contains about 650 feet of drifts and crosscuts. No. 3 adit is at an elevation of 3,225 feet, and is about 800 feet long. The Bridal Chamber level is at 3,295 feet elevation; and the Ella workings, from which the original high-grade silver ore was mined in 1885, is at about 3,375 feet elevation. The most recent mine workings is the Ophir adit, which is 1,050 feet northeast of the No. 1 adit portal, at an elevation of about 3,160 feet. Extensive work was undertaken in 1952 and 1953, in the Ophir adit, but the extent of the work is unknown to the writer. Judging from the size of the Ophir dump, several thousand feet of workings are present.

In the lower levels of the mine, which were accessible from the No. 1 adit, the ore bodies occurred in sheared and brecciated dolomitic limestone. The mineralized shear and breccia zones are roughly parallel, and within a few hundred feet of a northwest-trending fault that dips 50° to 60° SW. Less than 100 feet east of the main workings, the limestone is in contact with granite. Both the contact and the fault are almost barren of ore. The ore bodies of the sheared and brecciated limestone were irregular in shape and size, and appeared to have been controlled by intersecting fracture zones.

In the uppermost workings of the mine, which are known as the Ella workings, the ore bodies occur along gently dipping shear zones in limestone. The shear zones dip 10° to 20° NE. and are near a steeply dipping (80°), northeast-trending fault zone.

Ore minerals at the Old Dominion were chiefly galena and sphalerite, which were accompanied by minor argentite and native silver. Massive quartz, calcite, dolomite, and siderite were the main gangue minerals. In the Ella workings, the altered ore contained secondary cerussite and anglesite, as well as abundant limonite.

Principal Silver Deposits of Colville District

Longshot

Location: NW$\frac{1}{4}$ sec. 18, T. 36 N., R. 41 E.

Development: 600-foot crosscut adit with several hundred feet of drifts, raises, and stopes.

Geology: Replacement deposit in limy beds in argillite, phyllite, and quartzite. 20-foot bed sparsely metallized, and contained several ore shoots. Ore averaged 20.7 ozs. per ton in silver, 3.3 percent lead, and 4.2 percent zinc. Select samples assayed up to 150 ounces per ton in silver.

Ore minerals: Galena, sphalerite, tetrahedrite, and scheelite.

Production: 1951-1952: 246 tons of ore yielded 5,094 ozs. silver, 3 ozs.

gold, 16,330 lbs. lead, and 20,581
lbs. zinc.

References: Huntting, 1956, p. 244; Washington Division of Geology and Earth
Resources field notes.

Middleport

Location: SE¼ sec. 12, T. 36 N., R. 41 E.

Development: 60- and 800-foot adits.

Geology: ½- to 2-foot-thick quartz vein in
granodiorite. Ore averages 8 percent
lead, 15 percent zinc, 11.8 ozs. silver, and 0.13 oz. gold.

Ore minerals: Galena, sphalerite, chalcopyrite, tetrahedrite, and pyrite.

Production: Minor production in 1929; 1937-
1939.

References: Huntting, 1956, p. 379.

Old Dominion

Location: NW. cor. sec. 9, T. 35 N.
R. 40 E.

Development: About 40,000 feet of underground workings on at least 11 levels.
Main adit over 5,500 feet long.

Geology: Replacement deposits along shear
zones and in breccia in dolomitic
limestone. Ore averaged 83 ozs. per
ton in silver and 9 percent lead.
Near-surface ore averaged 400 ozs.
per ton in silver and 33 percent lead.

Ore minerals: Galena, sphalerite, argentite,
native silver, cerussite, and anglesite.

Production: 1885-1953: 6,588 tons of ore
yielded 323 ozs. gold, 942,517 ozs.
silver, 2,244,391 lbs. lead, 148,563
lbs. zinc, and 4,132 lbs. copper.

References: Weaver, 1920, p. 171-173;
Jenkins, 1924, p. 120-123; Huntting,

1956, p. 330; Bancroft, 1914, p. 126-
130.

Ore Cache

Location: Sec. 9, T. 35 N., R. 40 E.

Development: 40-foot shaft and several short
adits.

Geology: Replacement deposits in limestone.
Ore assayed 52 ozs. per ton in silver.

Ore minerals: Galena.

Production: 1916: 3 tons yielded 156 ozs.
silver and 640 lbs. lead.

References: Weaver, 1920, p. 173-174;
Huntting, 1956, p. 246.

NORTHPORT DISTRICT

Location

The Northport district is in northeastern
Stevens County, and lies between the Columbia River
on the west and the Pend Oreille-Stevens County
border on the east. Elevations range from 1,333 feet
at Northport to 7,200 feet at the summit of Abercrom-
bie Mountain. Most lead and zinc ores of the area
contain small amounts of silver; mines where the silver content of the ore exceeds 7 ounces per ton in
silver occur mainly in the extreme northeast corner
of the area (fig. 12).

Geology and Mineralization

Ore deposits of the Northport district are of
three general types: (1) replacement deposits along
shear zones in limestone and dolomite; (2) galena
in chimney deposits in dolomite; and (3) argentian
tetrahedrite in quartz veins in argillite. The richest
silver deposits are in the third type, whereas deposits

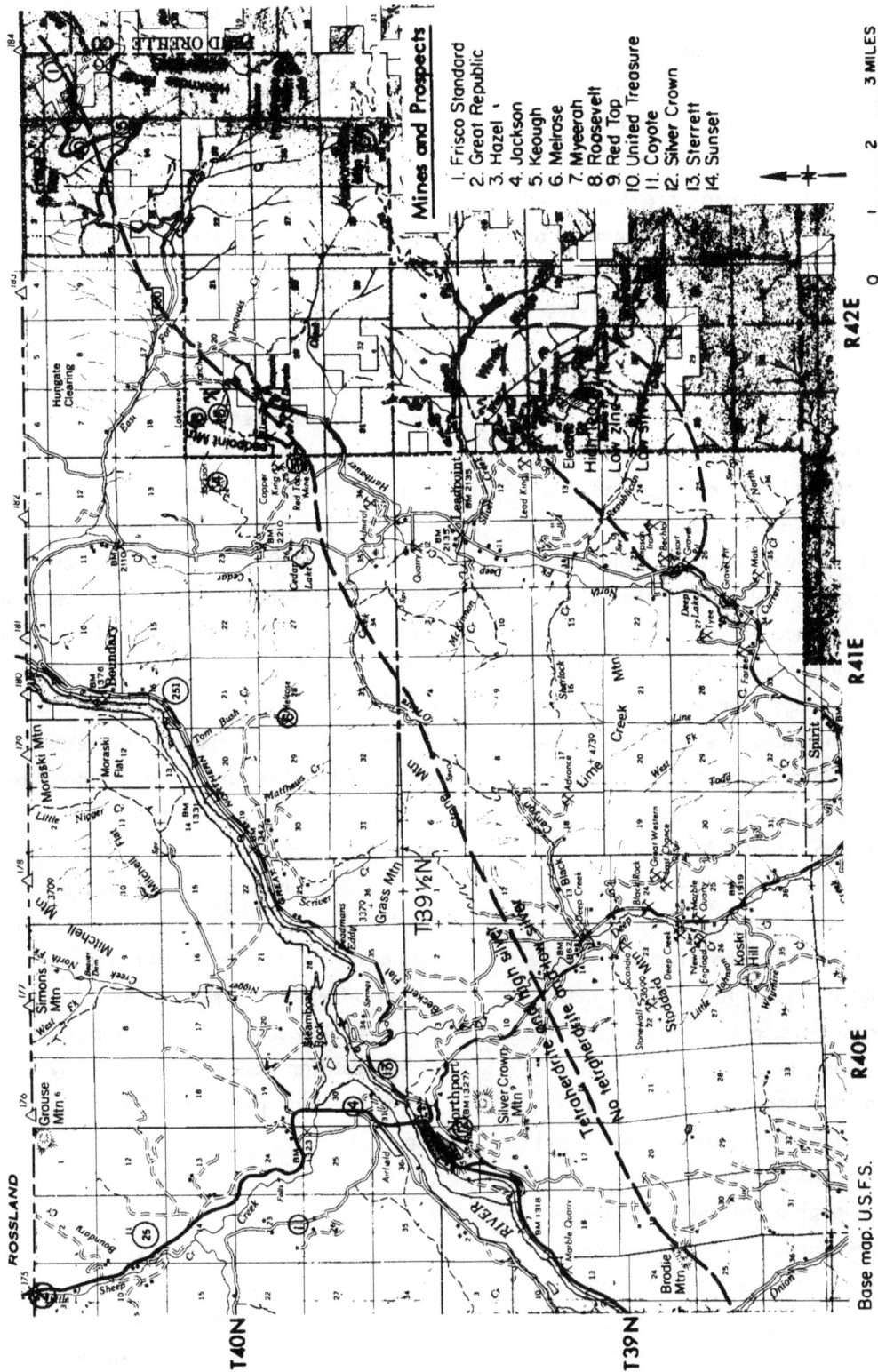

FIGURE 12—Index map of the Northport-Leadpoint district.

of the first type generally have the lowest silver content. However, the bulk of silver produced in the Northport area has come from deposits of the first type because a greater tonnage of ore has been mined from these deposits.

The predominate ore minerals are galena and sphalerite, which are almost always accompanied by pyrite, and occasionally by pyrrhotite. In addition to these minerals, several deposits contain varying amounts of tetrahedrite, chalcopyrite, scheelite, argentite, stannite, malachite, and azurite. The greatest assemblage of silver minerals occurs at the Frisco Standard mine where argentite, pyrargyrite, cerargyrite, and bromyrite have been reported. In general, ores rich in tetrahedrite are rich in silver. Although argentiferous galena is present in the Northport area, the silver content of most galena is low, and does not exceed 1 ounce per ton.

Deposits of high-silver contents are of the quartz vein type, which vary from several inches to as much as 14 feet in thickness; the average thickness is about 4 feet. Most veins of this type occupy shear zones or fissures in middle Paleozoic argillite and limestone that have been intruded by granitic rocks of the Spirit pluton (Cretaceous-Tertiary). Many of the veins have been sheared into small discontinuous lenses by postmineral faulting, and silver rich ore shoots seldom contained over 20 tons of high-grade silver ore. The richest silver ore mined from a quartz vein deposit came from the Jackson mine where 7 tons of ore averaged 188 ounces per ton in silver. The most productive quartz vein deposit was the Red Top from which 2,474 tons of zinc-lead-silver ore has been mined. Although argentiferous galena contained up to 16 ounces per ton in silver, the average silver content of ore mined was only 2.56 ounces per ton.

In the dolomite or limestone replacement deposits, galena, sphalerite, and pyrite are erratically distributed through the ore bodies. High-grade ore is separated from low-grade ore or nearly barren rock, but large-scale mining operations make it possible to mine the metallized rock at low costs. The silver content of the replacement deposits is generally low, with the average being around 0.02 ounce per ton. What little silver is present appears to be carried by the galena. The common host rock for lead-zinc replacement deposits of the Northport area is a middle dolomite member of the Metaline Limestone (Cambrian). The ore bodies appear to be controlled by bedding as well as faults.

In chimney deposits, which also occur in dolomite of the Metaline Limestone, galena forms nodules from a few inches to many feet in diameter. Most chimneys are 10 to 15 feet in diameter, 100 feet deep, and occur at brecciated intersections of shear zones. At the Electric Point mine, the largest chimney had a diameter of 150 feet, and was mined to a depth of 800 feet. At the Electric Point and Gladstone mines, the galena was accompanied by cerussite, anglesite, and abundant limonite. Ore shipped from these mines averaged 30 to 39 percent lead and 0.14 to 0.48 ounces per ton in silver.

According to Yates (1970, p. 22, fig. III-2), the Northport district can be divided mineralogically into areas of (1) tetrahedrite and high silver; (2) no tetrahedrite and low silver; and (3) high lead, low zinc, and low silver. Based on past production records, most significant silver mines are in the northeast part of the area of tetrahedrite and high silver (fig. 12).

Production

Silver produced in the Northport district came mainly from lead-zinc mining operations. From 1902 through 1969, 47 mines in the district produced around

141,000 tons of zinc, 47,000 tons of lead, 254,085 pounds of copper, 239,956 ounces of silver, and 982 ounces of gold. Only seven mines in the district are classed as silver mines. Ore from these mines averaged 7.3 to 200 ounces per ton in silver; however, total production from these mines was less than 1,000 tons.

Most silver produced in the Northport district has come from the Van Stone, Calhoun, Blue Ridge, and Deep Creek mines, which were major lead-zinc mines of Stevens County. These mines produced over 100,000 ounces of silver from ore that averaged only about 0.02 ounce per ton in silver.

After the closure of the Van Stone mine in 1970, silver production in the Northport district has been insignificant. Currently (1975), all major mines are idle, but exploration work in the district continues yearly.

Principal Silver Deposits of Northport District

Bullion
Location: Sec. 8, T. 39 N., R. 39 E.
Development: 160-foot shaft with 180 feet of drifts.
Geology: Quartz veins in schist and argillite.
Ore minerals: Galena, sphalerite, chalcopyrite, bornite, and pyrite.
Production: 11 tons in 1925 yielded 115 ozs. silver and 8,253 lbs. of lead.
References: Huntting, 1956, p. 238.

Burrus
Location: Center N$\frac{1}{2}$ sec. 8, T. 37 N. R. 41 E.
Development: 80-foot shaft and several short adits; underground workings totaling 300 feet.

Geology: Shear zone up to 60 feet wide in argillite contains small quartz veins. Ore averaged around 12 ozs. per ton in silver and 2 percent lead.
Ore minerals: Galena, chalcopyrite, and sphalerite.
Production: 1935, 1938, 1939: 81 tons yielded 1,026 ozs. silver, 3,155 lbs. lead, and 197 lbs. copper.
References: Huntting, 1956, p. 376; Colville Engineering Co., 1941, p. 91-92.

Coyote
Location: SW$\frac{1}{4}$ sec. 26 and NW$\frac{1}{4}$ sec. 35, T. 40 N., R. 39 E.
Development: Shallow shaft with short drift.
Geology: Metallized shear zones up to 2 feet thick in argillite and limestone.
Ore minerals: Argentiferous galena.
Production: 29 tons in 1916 and 1917 yielded yielded 156 ozs. silver and 3,696 lbs. lead.
References: Huntting, 1956, p. 239.

Frisco Standard
Location: S$\frac{1}{2}$NW$\frac{1}{4}$, sec. 12, T. 40 N., R. 42 E.
Development: Several caved adits with over 1,000 feet of workings; longest adit 500 feet.
Geology: Quartz veins or lenses up to 7 feet thick in sheared graphitic schist. Ore averaged around 8 ozs. per ton in silver, 1.15 percent lead, and 0.3 percent copper. High-grade ore contained 72 ozs. per ton in silver and 17 percent lead.
Ore minerals: Galena, tetrahedrite, chalcopyrite, sphalerite, pyrargyrite, cerargyrite, bromyrite, azurite,

malachite, and stannite.

Production: 1915-1944: 258 tons yielded
1 oz. gold, 2,044 ozs. silver, 1,586
lbs. copper, and 5,973 lbs. lead.

References: Patty, 1921, p. 112-114; Weaver,
1920, p. 304; Bancroft, 1914, p. 58-
59; Park and Cannon, 1943, p. 61-62.

Galena Farm

Location: NE¼ sec. 7, T. 37 N., R. 40 E.

Development: 300-foot adit and 2 shafts.

Geology: Metallized lenses of quartz and
calcite along shear zone in argillite
and dolomite. Assays of up to 9 ozs.
per ton in silver reported.

Ore minerals: Argentiferous galena, tetrahedrite, chalcopyrite, and pyrite.

Production: Unknown.

Reference: Huntting, 1956, p. 241.

Great Republic

Location: N¼ cor. sec. 3, T. 40 N.,
R. 39 E.

Development: 400-foot adit, 110-foot winze,
and 120-foot inclined shaft.

Geology: Two quartz-calcite veins up to 4
feet thick along diorite-limy shale
contact. Parts of ore vein contain
2.5 to 8.5 percent copper, 19 ozs.
silver, and up to 0.4 oz. gold per
ton.

Ore minerals: Chalcopyrite, galena, sphalerite, and pyrrhotite.

Production: None.

References: Bancroft, 1914, p. 60-61;
Colville Engineering Co., 1941,
p. 113-114; Weaver, 1920, p. 318.

Hazel

Location: Center sec. 19, T. 40 N.,
R. 42 E.

Development: 1,500 feet of adits.

Geology: Quartz vein in argillite ore averaged 32 ozs. per ton in silver, and
around 5 percent lead.

Ore minerals: Galena and silver sulfides.

Production: 1922: 24 tons yielded 770 ozs.
silver and 2,389 lbs. lead.

References: Jenkins, 1924, p. 102.

Jackson

Location: Center sec. 24, T. 40 N.,
R. 41 E.

Development: 350-foot adit with 100-foot
raise, and three adits each about
175 feet long.

Geology: Quartz vein 4 to 6 feet thick in
argillite. Ore averaged 188 ozs.
per ton in silver.

Ore minerals: Galena, sphalerite, chalcopyrite, and pyrite.

Production: 1934-1935: 7 tons yielded
1,318 ozs. silver and 181 lbs. lead.

References: Colville Engineering Co., 1941,
p. 108; Huntting, 1956, p. 378.

Keough

Location: NE¼ sec. 14, T. 40 N., R. 42 E.

Development: Short adit and open cuts.

Geology: Sparsely metallized quartz lenses
in dolomite. Ore averaged 10 ozs.
per ton in silver.

Ore minerals: Galena, sphalerite, and tetrahedrite.

Production: 1917: 14 tons yielded 1 oz. gold,
143 ozs. silver, and 3,260 lbs. copper.

References: Bancroft, 1914, p. 59

Melrose (Paragon)

Location: Near W¼ cor. sec. 38, T. 40 N.,
R. 41 E.

Development: 1,600 feet of underground
workings on two levels; 3 adits and
a 100-foot shaft.

Geology: 6-foot quartz vein in argillite contains scattered lenses and pods of ore
minerals. Small ore shoots contained
up to 100 ozs. per ton in silver.

Ore minerals: Tetrahedrite, galena, sphalerite, and pyrite.

Production: 1913-1938: 97 tons yielded
2,973 ozs. silver, 1,255 lbs. copper,
11,176 lbs. lead.

References: Patty, 1921, p. 116-117;
Colville Engineering Co., 1941,
p. 101; Huntting, 1956, p. 329.

Myeerah

Location: Center NE¼ sec. 11, T. 40 N.,
R. 42 E.

Development: 475 feet of adits and several
open cuts.

Geology: 2-foot-thick quartz vein in argillite and slate. Ore shoots contain up
to 20 ozs. per ton in silver, and average ore contained around 14 ozs. per
ton.

Ore minerals: Galena and tetrahedrite.

Production: 1926 and 1942: 21 tons yielded
288 ozs. silver and 8,715 lbs lead.

References: Park and Cannon, 1943, p. 62;
Colville Engineering Co., 1941,
p. 98.

Morning

Location: Center NW¼ sec. 4, T. 37 N.,
R. 41 E.

Development: 150-foot adit and several small
adits; drifts total about 400 feet.

·Geology: Quartz vein up to 4½ feet thick in
argillite and phyllite. Ore minerals
occur in lenses and pods up to 1 foot
thick. Ore averaged 90 ozs. per ton
in silver. Select samples contained
up to 350 ozs. of silver per ton.

Ore minerals: Galena, sphalerite, and pyrite.

Production: 1928-1948: 19 tons yielded
1,712 ozs. silver, 11 lbs. copper,
10,917 lbs. lead, and 4,898 lbs.
zinc.

References: Colville Engineering Co., 1941,
p. 89; Gage, 1941, p. 114.

Red Top

Location: NE¼SE¼ sec. 25, T. 40 N.,
R. 41 E.

Development: 2,700-foot adit with 1,500
feet of drifts, raises, and winzes.

Geology: Quartz veins in argillite and limestone. Ore shoots up to 4 feet thick
and 18 feet long averaged 2½ ozs. per
ton in silver and 4 percent lead.
High-grade ore contained up to 60
ozs. per ton in silver.

Ore minerals: Galena, sphalerite, chalcopyrite, pyrite, and scheelite.

Production: 1926-1956: 2,474 tons yielded

6,339 ozs. silver, 207,371 lbs. lead,
140,595 lbs. zinc, and 2,960 lbs.
copper.

References: Culver and Broughton, 1945,
p. 59-61; Huntting, 1956, p. 247.

Roosevelt

Location: Center N½ sec. 19, T. 40 N.,
R. 42 E.

Development: 800- and 400-foot adits.

Geology: 4-foot-thick quartz vein in phyllite
ore averaged 18 ozs. per ton in silver.

Ore minerals: Galena, sphalerite, and pyrite.

Production: 1937: 48 tons yielded 897 ozs.
silver and 1,004 lbs. lead.

References: Colville Engineering Co., 1941,
p. 109.

Silver Crown

Location: NE¼SE¼ sec. 5 and NW¼SW¼ sec.
4, T. 39 N., R. 40 E.

Development: 60-foot shaft and 130- and
100-foot adits.

Geology: Narrow quartz veins in limestone.
Veins contain up to 90 ozs. per ton in
silver.

Ore minerals: Galena, cerussite, sphalerite,
smithsonite, and pyrite.

Production: 34 tons in 1925 and 1926 yielded
309 ozs. silver and 5,674 lbs. lead.

Reference: Huntting, 1956, p. 247.

Sterrett

Location: Sec. 33, T. 40 N., R. 40 E.

Development: 500 feet of drifts and raises.

Geology: Veins up to 3 feet thick in limestone.

Ore minerals: Galena and minor sphalerite.

Production: 500 pounds shipped prior to 1941
assayed 56 percent lead, 3 percent
zinc, and 34 ozs. silver per ton.

References: Huntting, 1956, p. 248.

Sunset

Location: SW¼ sec. 30, T. 40 N., R. 40 E.

Development: 400-foot inclined shaft with
several levels.

Geology: Metallized contact up to several
feet thick at contact between diorite
and limestone. Contact assays 3 to
39 ozs. silver, 9 to 75 percent lead,
and up to 2 ozs. gold.

Ore minerals: Abundant pyrrhotite and minor
galena, sphalerite, tetrahedrite, and
chalcopyrite.

Production: Unknown.

Reference: Huntting, 1956, p. 248.

United Treasure

Location: Center sec. 11, T. 40 N.,
R. 42 E.

Development: 800-foot adit and 1,000 feet
of shallow adits and cuts.

Geology: Quartz vein up to 4 feet thick in
slate, and exposed for 400 feet along
its strike. Hand-sorted ore contained
117 ozs. per ton in silver. Average
ore contained 45 ozs. per ton in silver. Ore bodies were lenticular,
and sporadically distributed in the
vein.

Ore minerals: Tetrahedrite, galena, sphalerite, pyrite, malachite, and azurite.

Production: 1916-1953: 84 tons yielded
1 oz. gold, 3,783 ozs. silver, 30,702
lbs. lead, 695 lbs. copper, and 22

lbs. zinc.

References: Park and Cannon, 1943, p. 60;
Jenkins, 1924, p. 104-105; Patty,
1921, p. 114-116.

KETTLE FALLS DISTRICT

The Kettle Falls district of west-central Stevens County is a gold mining district, with minor production. From 1903 through 1953, mines of the district produced 9,192 tons of ore that yielded 1,479 ounces of gold, 40,137 ounces of silver, 61,974 pounds of copper, 32,368 pounds of lead, and 200 pounds of zinc. Although the average silver content of the ore produced in the district was but 5.06 ounces per ton, gold ore from the Gold Ledge mine contained as much as 37 ounces per ton in silver. Twenty-three mines in the district have a record of production, but only at two mines (Big Silver and Gold Ledge) did the value of silver exceed that of any other metal. Currently (1975) no mines are operating in the district.

The ore deposits of the Kettle Falls district occur in quartz fissure veins and along shear zones. Paleozoic limestone, argillite, and quartzite, as well as minor Tertiary intrusives, are common host rocks for the deposits. The quartz veins are from several inches to as much as 6 feet in thickness, whereas some metallized shear zones are as much as 30 feet wide. Some shear zones contain numerous quartz veins, which contain ore minerals, while in other shear zones the ore minerals are scattered throughout the sheared rock. Common ore minerals of the quartz vein deposits and shear zones are chalcopyrite, galena, sphalerite, and pyrite. The less common minerals include native gold and silver, stibnite, bornite, covellite, boulangarite, cuprite, molybdenite, malachite, and azurite. In most deposits the ore minerals occur as small discontinuous lenses or as sparsely scattered grains in the quartz veins and shear zones. Most ore shoots seldom contain over 100 tons of ore.

Principal Silver Deposits of Kettle Falls District

Aguila (Eagle-Newport)

Location: SE$\frac{1}{4}$ sec. 25, T. 35 N., R. 37 E.

Development: 580-foot adit with short cross-cuts, 65-foot shaft.

Geology: Numerous quartz veins in a 25-foot-wide shear zone in argillite and quartzite; parts of the shear zone are almost entirely quartz.

Ore minerals: Chalcopyrite, malachite, azurite, galena, tetrahedrite, and pyrite; 4 feet of vein assayed 18.5 ozs. silver, 1.5 percent copper, 6.67 percent lead.

Production: 1938: 1$\frac{1}{2}$ tons yielded 30 ozs. silver, 38 lbs. copper, 264 lbs. lead.

References: Huntting, 1956, p. 236; Washington Division of Geology and Earth Resources files.

Ark (Silver Queen)

Location: Center N$\frac{1}{2}$ sec. 11, T. 35 N., R. 37 E.

Development: 100- and 150-foot shafts with drifts, 1,700-foot adit with 130-foot raise.

Geology: Quartz lenses up to 4 feet thick in argillite and limestone. Mineralized zone up to 30 feet wide.

Ore minerals: Galena, sphalerite, tetrahedrite, chalcopyrite, and pyrite. Some ore shoots contained up to 24 ozs. silver.

Production: 1915-1949: 1,925 tons produced

in 1937, and 1938 yielded 27,754 ozs.
silver, 10,449 lbs. copper, 21,538
lbs. lead.

References: Huntting, 1956, p. 323; Patty,
1921, p. 143-144.

Gold Ledge

Location: SW¼ sec. 4, T. 36 N., R. 38 E.

Development: 140-foot shaft, 50-foot drift,
and 160-foot adit.

Geology: Quartz vein along contact between
argillite and porphyry.

Ore minerals: Free gold, silver sulfides,
galena, and pyrite. Ore averaged
36 ozs. per ton in silver.

Production: 1933-1935: 36 tons yielded
15 ozs. gold, 1,322 ozs. silver, 54
lbs. lead.

References: Huntting, 1956, p. 169; Fulkerson, 1958, p. 33.

Gold Reef (Benvenue)

Location: SW¼SE¼ sec. 9, T. 36 N., R. 38 E.

Development: 80-foot shaft, 200-foot drift,
and 700-foot adit.

Geology: 2- to 3-foot-thick quartz vein
along contact between argillite and
acidic dike.

Ore minerals: Free gold, silver sulfides, and
chalcopyrite. High-grade ore shoots
averaged 2.15 ozs. gold, and 8 ozs.
silver.

Production: 1905-1950: 704 tons yielded
709 ozs. gold, 3,200 ozs. silver.

References: Huntting, 1956, p. 169; Fulkerson, 1958, p. 33.

Vanasse

Location: S½NE¼ sec. 10, T. 36 N., R. 38 E.

Development: 8 shafts, 7 to 20 feet deep;
1 shaft, 65 feet deep.

Geology: 1- to 6-foot-thick quartz vein in
quartzite and arkose.

Ore minerals: Chalcopyrite, galena, sphalerite, tetrahedrite, stibnite, bornite,
and pyrite. Select samples assayed
as high as 14 percent lead, 6 percent
copper, and 140 ozs. per ton in silver.

Production: Unknown.

References: Huntting, 1956, p. 333; Purdy,
1951, p. 145.

ORIENT DISTRICT

The Orient district of northwestern Stevens
County is basically a gold district. From 1904 through
1942, 24 mines in the district produced 236,993 tons
of ore that yielded 45,057 ounces of gold, 32,020
ounces of silver, 555,536 pounds of copper, and
23,115 pounds of lead. The bulk of the gold and silver came from the First Thought mine near Orient.
From 1904 through 1942 the mine produced 38,436
ounces of gold and 16,699 ounces of silver from ore
that averaged 0.47 ounce per ton in gold, and 0.24
ounce per ton in silver. Almost all the copper produced in the district came from the Napoleon mine
near Boyds. Although several properties report up to
22 ounces per ton in silver, the average silver content of ore shipped from the district was less than 1
ounce per ton. Currently (1975) no properties are
operating in the district.

Ore deposits in the Orient district occur mainly
as replacement deposits in Paleozoic schist and amphibolite, and as metallized quartz veins and fracture
zones in latite and andesite of Mesozoic and early
Tertiary age. The deposits that are richest in silver

consist of quartz veins that range from 6 inches to 25 feet in thickness. Common ore minerals are pyrite, pyrrhotite, galena, sphalerite, and chalcopyrite. Bornite, tetrahedrite, and boulangerite are rarely present. The ore minerals are sparsely distributed or form narrow lenses and bands in most veins. To date, (1975), no large silver-rich ore shoots have been discovered in the district.

Principal Silver Deposits of Orient District

Comstock
Location: W$\frac{1}{2}$ sec. 25 and E$\frac{1}{2}$ sec. 26, T. 40 N., R. 36 E.

Development: Three shallow shafts and one adit.

Geology: Metallized shear zones and quartz veins 4 to 6 feet thick in quartzite and gneiss. One quartz vein contains up to 3.5 percent copper and 22 ozs. silver.

Ore minerals: Galena, chalcopyrite, and pyrite.

Production: Unknown.

Reference: Huntting, 1956, p. 95.

Easter Sunday
Location: Center E$\frac{1}{2}$ sec. 22, T. 40 N., R. 37 E.

Development: 130-foot inclined shaft, with levels at 70 and 100 feet.

Geology: 2- to 6-foot-thick quartz vein in cherty argillite.

Ore minerals: Pyrite, chalcopyrite, sphalerite, tetrahedrite, and boulangerite.

Production: 20 ton shipment in 1909 averaged

1.21 percent copper, 12.4 ozs. silver, and 0.44 oz. gold per ton.

References: Huntting, 1956, p. 168.

Iron Mask
Location: Sec. 25, T. 40 N., R. 36 E.

Development: 3 shafts and several open cuts.

Geology: 4- to 6-foot thick quartz vein in latite and limestone. Vein traceable for 3,000 feet. Select sample of vein assayed 7 ozs. silver, 0.08 oz. gold, 14 percent lead, and 28 percent zinc.

Ore minerals: Sphalerite, galena, chalcopyrite, pyrite, and pyrrhotite.

Production: None.

Reference: Huntting, 1956, p. 378.

McKinley
Location: Sec. 34, T. 39 N., R. 38 E.

Development: Unknown.

Geology: 4-foot "ledge" in andesite. Select samples assay up to 200 ozs. per ton in silver.

Ore minerals: Galena and pyrite.

Production: None.

Reference: Huntting, 1956, p. 244.

Montana and Washington
Location: Sec. 30, T. 40 N., R. 37 E.

Development: 215-foot adit.

Geology: Quartz veins 6 inches to 25 feet thick in gneiss. Select samples assayed 16 to 20 ounces per ton in silver.

Ore minerals: Chalcopyrite, bornite, galena, and pyrite.

Production: Unknown.

Reference: Huntting, 1956, p. 100.

FERRY COUNTY

Ferry County is Washington's leading silver-producing county, having produced 5,917,851 ounces of the metal from 1904 through 1956. Since 1956 the county has produced over 100,000 ounces of silver yearly, almost all of which came from gold-silver mining operations in the Republic district.

Silver has been reported at 49 mines, and at 134 prospects in the county. The distribution of silver according to the major metal at each property is shown in table 21.

TABLE 21.—Distribution of silver in Ferry County

Major metal at property	Mines	Prospects	No. of mines or prospects
Lead	3	53	56
Gold	28	24	52
Silver	12	38	50
Copper	5	16	21
Zinc	0	2	2
Nickel	1	1	2

Although 49 mines have produced silver along with other metals, 95 percent of the silver mined in Ferry County has come from gold mines of the Republic district. Production of silver from 12 silver mines in the county has been insignificant, and did not exceed 2,000 ounces.

Deposits that contain silver in Ferry County occur mainly in the Republic, Keller, and Covada districts (fig. 13). Silver in the Republic district is commonly associated with gold; in the Keller district, silver accompanies gold and copper; in the Covada district, silver is found in deposits of lead, zinc, copper and antimony. Silver is also present in the Belcher, Danville, Orient, and Sheridan districts where it is associated with gold and copper.

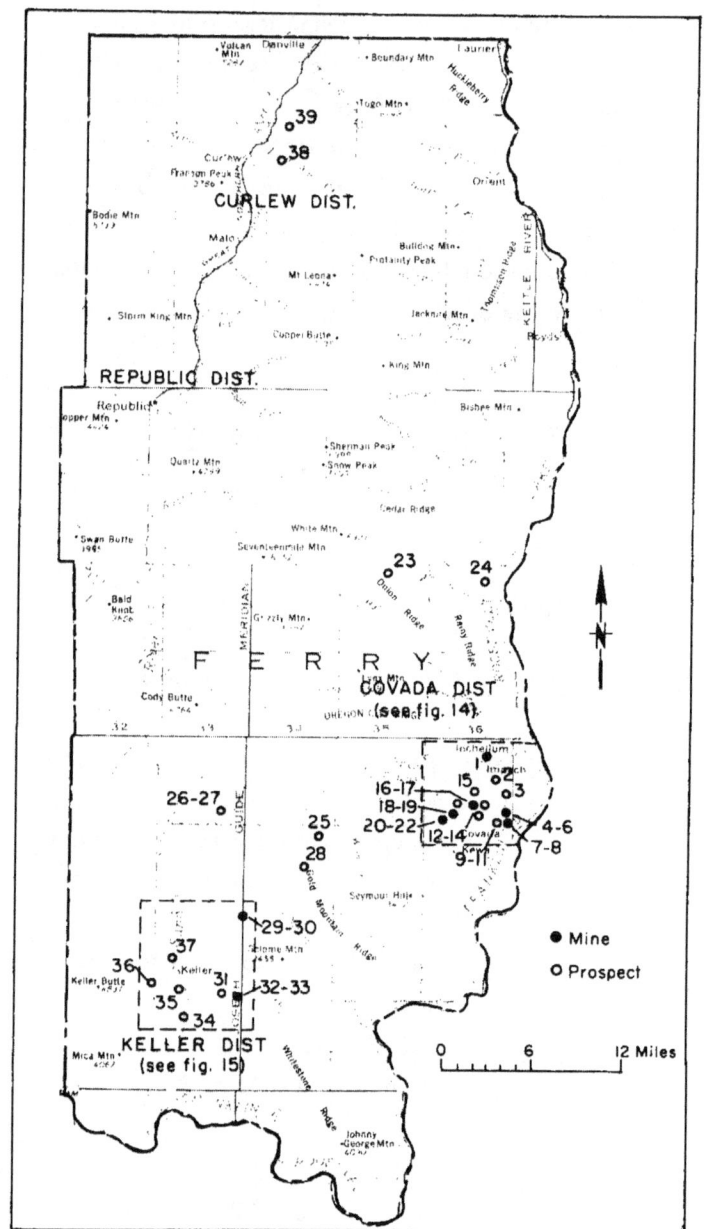

Base from U.S.G.S. State of Washington 1:500,000, 1962

Properties

1. Gwin	14. Royal Ann	27. Congress
2. Dan Patch	15. Good Ore	28. Burlington and Delaware
3. Algonkian	16. Stray Dog	
4. Longstreet	17. Oom Paul	29. Shamrock
5. Silver Leaf	18. Montana	30. Anderson
6. Shoo Fly	19. Romulus	31. Summit
7. Silver Crown	20. New York	32. Addison
8. Ivanhoe	21. Meteor	33. Golden Chord
9. Butterfly	22. Reserve	34. Jumper
10. King Solomon	23. Chief Barnaby	35. Rover Bonanza
11. Plymouth Rock	24. Big Lake	36. Meadow Creek
12. Advance	25. U.S.	37. Great Northern
13. Keystone	26. McJunkin	38. Anchor
		39. Panama

FIGURE 13.—Silver deposits of Ferry County.

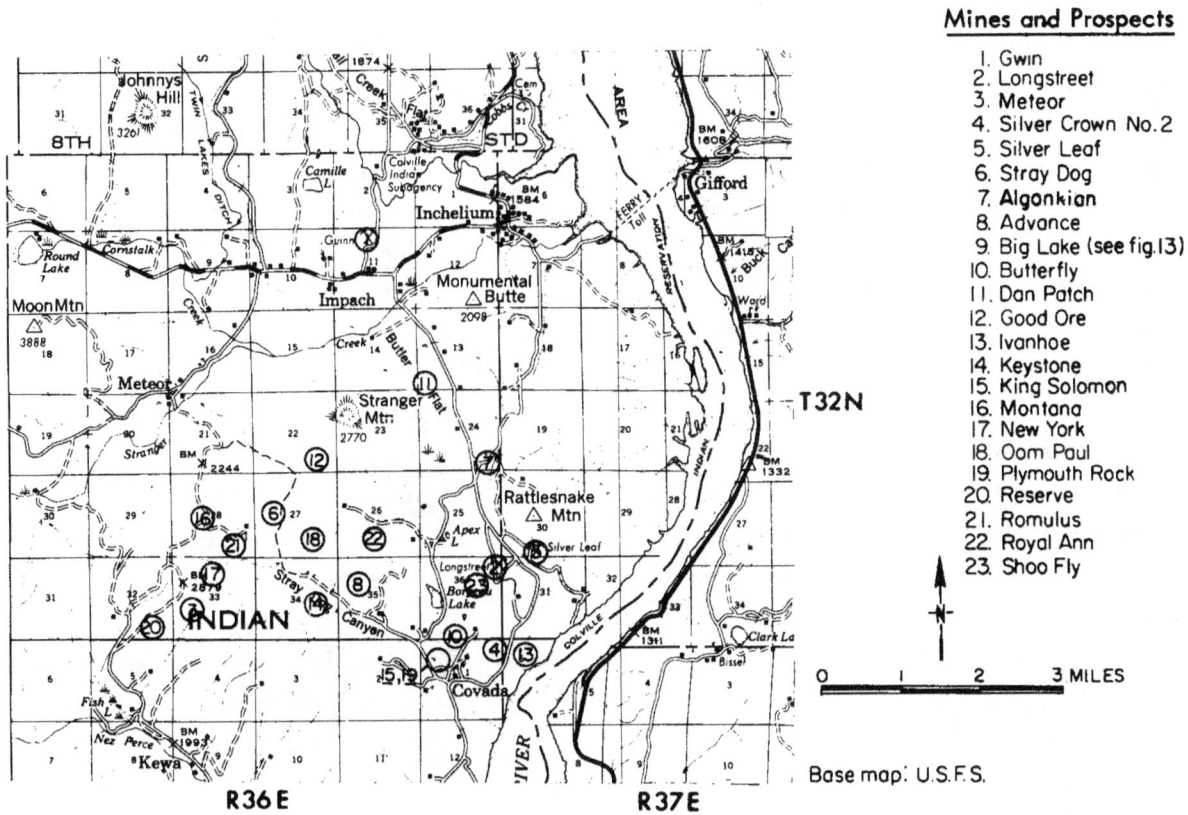

Mines and Prospects

1. Gwin
2. Longstreet
3. Meteor
4. Silver Crown No. 2
5. Silver Leaf
6. Stray Dog
7. Algonkian
8. Advance
9. Big Lake (see fig.13)
10. Butterfly
11. Dan Patch
12. Good Ore
13. Ivanhoe
14. Keystone
15. King Solomon
16. Montana
17. New York
18. Oom Paul
19. Plymouth Rock
20. Reserve
21. Romulus
22. Royal Ann
23. Shoo Fly

FIGURE 14.—Index map of the Covada district.

COVADA DISTRICT

Location

The Covada district is in southeastern Ferry County, immediately west of Lake Roosevelt. The district is entirely within the boundaries of the Colville Indian Reservation and centers about Twin Lakes. The district includes the old mining camps of Meteor, Impach, and Covada, and the present town of Inchelium. Elevations range from around 1,300 feet on the shores of Lake Roosevelt to around 6,000 feet on mountain peaks in the west part of the district. The silver-bearing deposits of the district occur mainly south of Inchelium in a northeast-trending area that is about 4 miles wide and 5 miles long (fig. 14).

Geology and Mineralization

Most ore deposits of the Covada district occur in quartz fissure veins, which have graywacke, argillite, and impure quartzite and limestone as their host rocks. All rocks are part of the Covada Group (Paleozoic), which are underlain by Cretaceous granodiorite. Along parts of the intrusive contact between the Covada Group and the granodiorite, hornfels has developed, some of which is sparsely metallized.

Common silver minerals of the quartz veins are argentiferous galena, pyrargyrite, cerargyrite, and argentite. Other minerals of the veins include galena, sphalerite, chalcopyrite, tetrahedrite, wolframite, scheelite, molybdenite, and stibnite. Pyrite is common in the veins; near the surface it has oxidized to limonite, which gives the veins a reddish-brown color. In addition to pyrite, some veins contain pyrrhotite and minor arsenopyrite.

Most veins strike NW. and dip 45° to 70° NE. They vary from several inches to as much as 10 feet in thickness; however, the average thickness of 50 veins is only 2 feet. Shafts have been sunk on several veins to a depth of 150 feet, and some veins increase in thickness with depth. Most ore shoots are lenticular, and are only several inches thick. Around 100 tons of ore was extracted from the longest ore bodies. Weaver (1913, p. 44-84) reports that the average silver content of the veins is 2.21 ounces. The average ore shipped from the district contained 45 ounces per ton in silver, while one small shipment from the Silver Leaf mine contained 180 ounces per ton in silver.

Production

Twelve mines in the Covada district have a record of production; however, the combined silver production from the mines was only around 10,000 ounces. Silver was the major metal at nine mines, whereas, lead was the major metal at three mines. The average mine in the district only produced about 50 tons of ore. In addition to silver, the mines produced around 30 ounces of gold, 5,600 pounds of lead and 500 pounds of copper.

Principal Silver Deposits of Covada District

Advance

Location: Center NW¼ sec. 35, T. 32 N., R. 36 E.
Development: 1,861 feet of workings in cross-cuts, drifts, raises, and winzes.
Geology: 2- to 3-foot-wide shear zone in schist contains quartz lenses up to 16 inches thick. Some quartz assays

0.05 to 333 ozs. silver, trace to 0.93 oz. gold, and up to 8 percent lead.
Ore minerals: Galena, pyrargyrite, native silver, chalcopyrite, sphalerite, and pyrite.
Production: Unknown.
Reference: Huntting, 1956, p. 288.

Algonkian

Location: SE¼ sec. 24, T. 32 N., R. 36 E.
Development: 8-foot shaft and open cut.
Geology: 6-inch-thick quartz vein in quartzite. Parts of the vein contain 6.25 to 8 ozs. silver and 0.03 to 0.06 oz. gold.
Ore minerals: Finely divided sulfides.
Production: Unknown.
Reference: Huntting, 1956, p. 288.

Big Lake

Location: Secs. 2 and 11, T. 36 N., R. 36 E.
Development: 793-foot adit; 20-, 30-, and 142-foot shafts.
Geology: 3-foot-wide shear zone in schistose grit and quartzite. Shear zone contains small quartz veins, parts of which assay up to 15 percent lead, 3 percent zinc, 40.32 ozs. silver, and 0.44 oz. gold.
Ore minerals: Galena, sphalerite, chalcopyrite, and pyrite.
Production: Unknown.
Reference: Huntting, 1956, p. 208.

Butterfly

Location: S¼ cor. sec. 36, T. 32 N., R. 36 E.
Development: 200-foot adit and 13-foot shaft.

Geology: 4- to 12-inch-thick quartz vein
in granodiorite. Select sample of
vein contained 30 ozs. per ton in
silver.

Ore minerals: Argentiferous galena.

Production: Unknown.

Reference: Huntting, 1956, p. 209.

Chief Barnaby

Location: SW¼ sec. 35, T. 35 N., R. 35 E.

Development: 395-foot adit and a 35-foot
shaft.

Geology: Shear zone in argillite, limestone,
and granodiorite. Shear zone contains lenses of ore minerals. High-
grade ore contained 21 percent lead
and 205 ozs. of silver per ton.

Ore minerals: Galena, pyrite, and sphalerite.

Production: Unknown.

Reference: Huntting, 1956, p. 289.

Dan Patch

Location: Near center SW¼ sec. 13,
T. 32 N., R. 36 E.

Development: 54-foot adit and a 30-foot
shaft.

Geology: Quartz veins up to 16 inches thick
in quartzite and slate. Vein assays
10.15 ozs. silver and 0.04 oz. gold.

Ore minerals: Sparse galena, sphalerite, and
pyrite.

Production: Unknown.

Reference: Huntting, 1956, p. 289.

Good Ore

Location: SE¼ sec. 22, T. 32 N., R. 36 E.

Development: 140-foot crosscut adit and a
10-foot shaft.

Geology: 6- to 10-inch-thick quartz vein
in granodiorite. Ore minerals occur
in thin bands which assay 9 to 89 ozs.
silver and 0.6 to 4.06 ozs. gold.

Ore minerals: Galena and pyrite.

Production: Unknown.

Reference: Huntting, 1956, p. 290.

Gwin

Location: NW¼ sec. 11, T. 32 N., R. 36 E.

Development: 750-foot adit and an inclined
shaft.

Geology: Quartz veins up to 4 feet thick in
quartzite and argillite. Vein averaged
27.7 ozs. in silver, 0.049 oz. in
gold, and 3 percent copper.

Ore minerals: Tetrahedrite, galena, wolframite, tennantite or enargite, pyrite,
and arsenopyrite.

Production: Mainly silver in the early 1900's,
but some tungsten during World War I.

References: Culver and Broughton, 1945
p. 17-19; Pardee, 1918, p. 173-174;
Weaver, 1913, p. 81-83.

Ivanhoe

Location: N½NW¼ sec. 6, T. 31 N.,
R. 37 E.

Development: 650-foot crosscut adit.

Geology: 3 quartz veins in calcareous shale.
Veins average 10 inches in thickness
and contain fair amounts of silver.

Ore minerals: Unknown.

Production: Unknown.

Reference: Huntting, 1956, p. 291.

Keystone

Location: Center E½ sec. 34, T. 32 N.,
R. 36 E.

Development: 550-foot adit.

Geology: 2- to 6-foot-wide metallized shear
zone in schist and carbonaceous shale.
Quartz veinlets in the shear zone
assay up to 190 ozs. silver and 4 ozs.
gold.

Ore minerals: Galena, native silver, pyrrhotite, sphalerite, chalcopyrite, and
pyrite.

Production: Unknown.

Reference: Huntting, 1956, p. 291.

King Solomon

Location: NW¼ sec. 1, T. 31 N., R. 36 E.

Development: Several shafts and adits.

Geology: 3 quartz veins ranging from 4 to
12 inches thick in granodiorite. One
vein averages 7.8 ozs. in silver per
ton.

Ore minerals: Galena and pyrite.

Production: None.

Reference: Huntting, 1956, p. 211.

Longstreet

Location: NE. cor. sec. 36, T. 32 N.,
R. 36 E.

Development: 400- and 80-foot adits.

Geology: 8- to 30-foot-wide metallized
granodiorite breccia. Ore contained
15 to 16 ozs. per ton in silver; high-
grade ore contained 144 ozs. per ton
in silver, and 24 percent antimony.

Ore minerals: Stibnite, argentite, galena,
tetrahedrite, sphalerite, chalcopyrite,
and pyrite.

Production: Several hundred tons in the early
1900's.

References: Purdy, 1951, p. 68-71; Pardee,
1918, p. 167-168.

Meteor

Location: SW¼ sec. 33, T. 32 N., R. 36 E.

Development: 700-foot crosscut with a 125-
foot drift and several stopes; 160-foot
inclined shaft.

Geology: 1- to 1½-foot-thick quartz vein in
carbonaceous argillite, quartzite,
and schist. High-grade ore shoots
contained 65 to 178 ozs. per ton in
silver, and ½ to 3½ percent lead.

Ore minerals: Galena, sphalerite, chalcopyrite, cerargyrite, pyrargyrite,
argentite, tetrahedrite, native silver,
pyrite, and rhodochrosite.

Production: 20 tons of ore by 1918.

References: Pardee, 1918, p. 155-156;
Huntting, 1956, p. 292.

Montana

Location: Center of sec. 28, T. 32 N.,
R. 36 E.

Development: 3 shafts from 20 to 90 feet deep.

Geology: 1-foot-thick quartz vein in granodiorite. Select samples contained up
to 47.6 percent lead and 224.8 ozs.
of silver.

Ore minerals: Minor galena, sphalerite,
pyrargyrite, and pyrite.

Production: Has produced, but amount unknown.

Reference: Huntting, 1956, p. 292.

New York

Location: Center N½ sec. 33, T. 32 N.,
    R. 36 E.

Development: 40- and 60-foot shafts.

Geology: 12- to 16-inch-thick quartz vein
    in granodiorite. Parts of the vein
    contain 15 percent lead, 10 percent
    zinc, 7.8 ozs. silver, and 0.07 oz.
    gold.

Ore minerals: Galena, pyrargyrite, sphal-
    erite, pyrite, and arsenopyrite.

Production: None.

Reference: Huntting, 1956, p. 212.

Oom Paul

Location: Center SE¼ sec. 27, T. 32 N.,
    R. 36 E.

Development: 60-foot adit and a 50-foot
    shaft.

Geology: 4- to 12-inch-thick quartz vein in
    granodiorite. Ore minerals make up
    50 percent of some parts of the vein.
    Select samples assay 69 percent lead
    and 25 ozs. silver.

Ore minerals: Galena and pyrite.

Production: None.

References: Huntting, 1956, p. 212.

Plymouth Rock

Location: N½NW¼ sec. 1, T. 31 N.,
    R. 36 E.

Development: Unknown.

Geology: Quartz vein up to 2½ feet thick at
    contact between limestone and grano-
    diorite. Vein assays up to 6 percent

lead and 15 ozs. silver.

Ore minerals: Galena, sphalerite, and pyrite.

Production: None.

Reference: Huntting, 1956, p. 212.

Reserve

Location: SE¼ sec. 32, T. 32 N., R. 36 E.

Development: 48-foot shaft with 75-foot
    drift at bottom.

Geology: 3-foot-thick quartz vein in a 7-
    foot-wide shear zone in quartzite.
    Vein averages 7.8 ozs. silver and
    0.04 oz. gold.

Ore minerals: Galena and pyrite.

Production: None.

Reference: Huntting, 1956, p. 293.

Silver Crown No. 2

Location: Near NE. cor sec. 1, T. 31 N.,
    R. 36 E.

Development: 112-foot shaft with drifts on
    50- and 100-foot levels; 125-foot
    crosscut.

Geology: 1½- to 4-feet-thick quartz vein in
    schistose quartzite and limy shale.
    Parts of vein containing fine-grained
    pyrite and unidentified sulfides assayed
    around 15 ozs. per ton in silver, and
    0.05 oz. in gold.

Ore minerals: Unidentified sulfides and
    pyrite.

Production: 1,935 lbs. in 1905 netted $168.

References: Pardee, 1918, p. 166; Weaver,
    1913, p. 57-58.

Silver Leaf

Location: Near S¼ cor. sec. 30, T. 32 N. R. 37 E.

Development: 50- and 240-foot adits;
40-foot shaft.

Geology: Quartz lenses in 50-foot-wide
shear zone in limestone, schist, and
argillite. Hand-sorted ore contained
180 ozs. per ton in silver, 3.6 percent lead, and 2.2 percent zinc.

Ore minerals: Tetrahedrite, chalcopyrite,
sphalerite, galena, stibnite, scheelite,
native silver, pyrargyrite, wolframite,
and pyrite.

Production: 1911, 1912, 1915, 1916, and
1935. Total tonnage small; probably
2,000 ounces of silver produced.

References: Pardee, 1918, p. 170-171;
Culver and Broughton, 1945, p. 19-
21; Purdy, 1951, p. 71-74; Gage,
1941, p. 117-118.

Shoo Fly

Location: Center NE$\frac{1}{4}$ sec. 36, T. 32 N.,
R. 36 E.

Development: 165-foot adit and a 6-foot
shaft.

Geology: 1$\frac{1}{2}$- to 4-foot-thick quartz vein
in granodiorite. Also an 80-foot
zone of metallized aplite.

Ore minerals: Galena and pyrite.

Production: 12 tons mined in 1967 yielded
105 ozs. silver and 900 lbs. lead.

References: Huntting, 1956, p. 213.

Stray Dog

Location: Near center W$\frac{1}{2}$ sec. 27, T. 32 N.
R. 36 E.

Development: 3 adits with 983 feet of workings, and a 100-foot shaft. Most

work undertaken in lowest adit, which
contains a drift, raise, and winze.

Geology: Quartz vein up to 2 feet thick in
shear zone in quartz monzonite porphyry. Parts of vein contain 5 to 66
ozs. per ton in silver, and 0.3 to
0.6 oz. gold.

Ore minerals: Chalcopyrite, galena, sphalerite, tetrahedrite, pyrargyrite,
argentite, native silver, pyrite, and
arsenopyrite.

Production: 40 tons prior to 1910.

References: Gage, 1941, p. 123-124,
Pardee, 1918, p. 159-160; Huntting,
1956, p. 294.

Romulus

Location: SE. cor. sec. 28, T. 32 N.,
R. 36 E.

Development: Caved incline shaft 110 feet
deep.

Geology: Closely spaced quartz veinlets in
a 5-foot-wide shear zone in granodiorite. Quartz assays up to 8 ozs. per
ton in silver.

Ore minerals: Galena and pyrite.

Production: None.

Reference: Huntting, 1956, p. 213.

Royal Ann

Location: Center S$\frac{1}{2}$ sec. 26, T. 32 N.,
R. 36 E.

Development: 3 short adits and 2 shallow
shafts.

Geology: Quartz veins up to 16 inches
thick in granodiorite. Veins assay
up to 12.5 ozs. silver.

Ore minerals: Galena, sphalerite, cerussite,
      and pyrite.

Production: None.

Reference: Huntting, 1956, p. 213.

U.S.

Location: Center W$\frac{1}{2}$ sec. 1, T. 31 N.,
      R. 34 E.

Development: 50-foot crosscut adit with a
      30-foot drift.

Geology: 6- to 10-inch quartz vein along
      a schist-limestone contact. Vein
      contains a 2-inch band of almost
      pure galena, which assays 49.5
      percent lead and 59.96 ozs. of silver.

Ore minerals: Galena and pyrite.

Production: None.

Reference: Huntting, 1956, p. 214.

KELLER DISTRICT

Location

The Keller district is in southwestern Ferry
County and falls within the boundaries of the Colville
Indian Reservation. The San Poil River flows through
the center of the district, and Lake Roosevelt forms
the district's southern boundary. The predominant
metal of the district is copper, but gold, silver, lead,
zinc, nickel, tungsten, and molybdenum are also
present. Silver has been reported at 5 mines and at
29 prospects; however, only at 3 mines and at 8 prospects is it present in significant amounts.

Geology and Mineralization

The ore deposits occur as metallized quartz
veins and shear zones in schist, limestone, argillite,

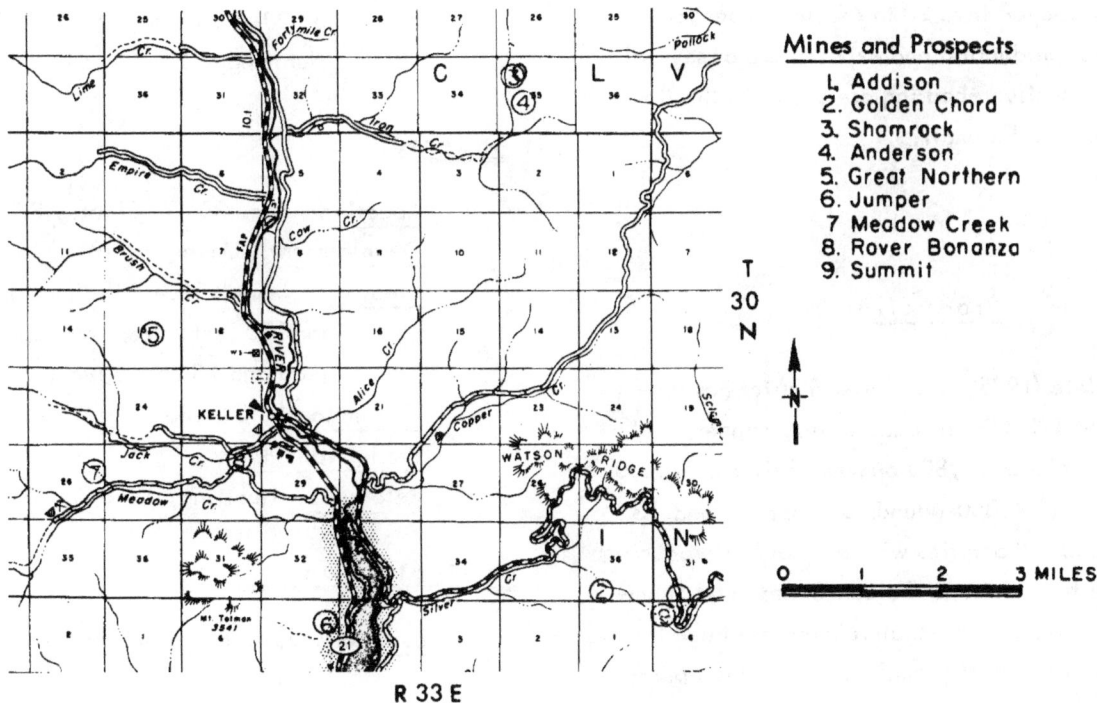

Mines and Prospects
1. Addison
2. Golden Chord
3. Shamrock
4. Anderson
5. Great Northern
6. Jumper
7 Meadow Creek
8. Rover Bonanza
9. Summit

FIGURE 15.—Index map of the Keller district.

quartzite, and serpentine of the Covada Group (Paleozoic). The rocks occur as isolated roof pendants that are surrounded by Cretaceous granodiorite of the Colville batholith. The quartz veins are up to 6 feet thick, whereas, the shear zones are as much as 200 feet wide. In most veins and shear zones, ore minerals are sparse, but in parts of some veins, the ore minerals form pods and lenses up to several feet thick that are of ore grade.

The predominant ore minerals are chalcopyrite, galena, and sphalerite. Other ore minerals that are less common include chalcocite, scheelite, wolframite, cerussite, nickeliferous pyrite, and molybdenite. Pyrite, and occasionally arsenopyrite, almost always accompanies the ore minerals. Silver minerals have not been identified in any of the deposits, but some galena contains up to 100 ounces of silver per ton. Select samples from ore shoots at eight prospects are reported to contain 5.5 to 170 ounces per ton in silver. Small shipments of ore from three mines assayed from 20 to 64 ounces per ton. Based on past production and exposed ore at several properties most silver-bearing ore shoots in the Keller district appear to be small.

Production

To date (1975), the Keller district has produced around $12,000 in lead, silver, copper, and gold. This includes 13,800 ounces of silver, 51,000 pounds of lead, 4,300 pounds of copper, and 13 ounces of gold. Properties with a record of production include the Addison, Golden Chord, and Shamrock mines. The most productive mine has been the Addison, where currently small-scale mining operations are underway.

Principal Silver Deposits of Keller District

Addison (Pacific Mutual)

Location: SW$\frac{1}{4}$ sec. 31, T. 30 N., R. 34 E.

Development: About 1,700 feet of underground workings in two adits and one shaft.

Geology: 4- to 6-foot-thick quartz veins in schist. Ore shoots up to 100-feet long assayed 1.4 to 33.8 ounces per ton in silver, up to 3.8 percent copper, and up to 19.6 percent lead.

Ore minerals: Galena, chalcopyrite, chalcocite, sphalerite, scheelite, wolframite, and pyrite.

Production: Minor production in 1923; 1,500 tons in 1973.

References: Culver and Broughton, 1945, p. 23-24; Gage, 1941, p. 142-143; Huntting, 1956, p. 51.

Anderson

Location: Sec. 35, T. 31 N., R. 33 N.

Development: Unknown.

Geology: 1-foot "vein" in limestone exposed for 150 feet. Vein averages 4 percent lead and 15 ozs. silver.

Ore mineral: Galena.

Production: None.

Reference: Huntting, 1956, p. 208.

Burlington and Delaware

Location: S$\frac{1}{2}$ sec. 15 and N$\frac{1}{2}$ sec. 22, T. 31 N., R. 34 E.

Development: 270-foot adit.

Geology: 5-foot-thick vein of quartz and
limonite at contact between limestone
and argillite. Parts of vein contain
16 ozs. silver and 5 percent lead.
Ore minerals: Galena, cerussite, and pyrite.
Production: None.
Reference: Huntting, 1956, p. 209.

Congress

Location: NW¼SE¼ sec. 35, T. 32 N.,
R. 33 E.
Development: 3 adits with a total of around
2,000 feet of workings.
Geology: Metallized shear zone up to 40
feet wide in serpentine at schist contact. Shear zone contains 0.17 to
5.17 percent nickel, 0.013 to 0.35
percent cobalt, and up to 5.5 ozs.
silver.
Ore minerals: Nickeliferous pyrite, chalcopyrite, malachite, and garnierite in
gangue of quartz, dolomite, barite,
and epidote.
Production: Unknown.
Reference: Huntting, 1956, p. 276.

Golden Chord

Location: SW¼SW¼ sec. 36, T. 30 N.,
R. 33 E.
Development: 125- and 400-foot adits, connected by raise.
Geology: Quartz veinlets in hornfels and
quartz monazite porphyry. Ore, at
shallow depth, contained 2 to 6 percent copper, and 2.4 to 64.2 ounces
per ton in silver. Ore occurs in small
pods and sparsely disseminated grains.

Ore minerals: Chalcopyrite, sphalerite,
galena, and pyrite.
Production: Minor shipments in the early
1900's.
References: Pardee, 1918, p. 132-133;
Huntting, 1956, p. 290.

Great Northern

Location: Near north line SE¼ sec. 13,
T. 30 N., R. 32 E.
Development: Open cuts.
Geology: 1-foot-thick quartz vein in grano-
diorite. Select samples from vein
contained 170 ozs. per ton in silver.
Ore minerals: Galena, sphalerite, and
pyrite.
Production: None.
Reference: Huntting, 1956, p. 290.

Jumper

Location: Center NE¼ sec. 5, T. 29 N.,
R. 33 E.
Development: Open cut.
Geology: Closely spaced quartz veinlets in
a 3-foot-wide shear zone in grano-
diorite. Select ore samples contained
7.5 percent copper, and 11.4 ozs.
silver.
Ore minerals: Chalcopyrite, molybdenite,
malachite, and pyrite.
Production: None.
Reference: Huntting, 1956, p. 53.

McJunkin

Location: Center NE¼ sec. 27, T. 32 N.,
R. 33 E.

Development: 45-foot adit with a 15-foot
    crosscut.
Geology: 4-foot-wide breccia zone between
    porphyry and limestone. Select sam-
    ples from breccia contained 23.36 ozs.
    silver, 7.7 percent lead, and consid-
    erable zinc.
Ore minerals: Galena, sphalerite, garnierite,
    and pyrite.
Production: None.
Reference: Huntting, 1956, p. 291.

Meadow Creek

Location: Secs. 25 and 26, T. 30 N.,
    R. 32 E.
Development: 8 adits with 2,103 feet of
    workings.
Geology: Quartz veins and stringers along
    shear zone in granodiorite. Parts of
    some veins contained 5 percent lead,
    4 percent copper, and 27 ozs. silver.
Ore minerals: Chalcopyrite, chalcocite,
    molybdenite, sphalerite, and galena.
Production: Unknown.
Reference: Huntting, 1956, p. 54.

Rover Bonanza

Location: Center NE$\frac{1}{4}$ sec. 30, T. 30 N.,
    R. 33 E.
Development: 4 adits from 120 to 400 feet
    long.
Geology: Quartz veins up to 18 inches thick
    in sheared granodiorite. Parts of one
    vein contained 52.9 ozs. silver and
    0.09 oz. gold; also high in lead and
    zinc.

Ore minerals: Galena, sphalerite, pyrite,
    and arsenopyrite.
Production: None.
Reference: Huntting, 1956, p. 293.

Shamrock (Iron Creek)

Location: SW$\frac{1}{4}$NW$\frac{1}{4}$ sec. 35, T. 31 N.,
    R. 33 E.
Development: Over 4,000 feet of underground
    workings in 3 adits and 2 shafts.
Geology: Metallized shear zone in silicified
    limestone. Ore averaged 30 ozs. of
    silver per ton, and 5 percent lead.
    Select samples of ore contained 1,000
    ozs. per ton in silver, and 75 percent
    lead.
Ore minerals: Galena, cerussite, chalcopyrite,
    and pyrite.
Production: 1914, 1922 and 1926: total sil-
    ver production around 7,000 ounces.
References: Huntting, 1956, p. 276; Patty,
    1921, p. 217-218.

Summit

Location: SW. cor. sec. 31, T. 30 N.,
    R. 34 E.
Development: 2 shafts.
Geology: 1-foot-thick quartz vein in argil-
    lite and quartzite. Select dump sam-
    ples assayed 14.2 percent zinc, 5.4
    percent copper, and 17.7 ozs. silver.
Ore minerals: Sphalerite with minor galena,
    pyrite, chalcopyrite, and specularite.
Production: None.
Reference: Huntting, 1956, p. 362.

CURLEW DISTRICT

The Curlew district of northern Ferry County
is basically a copper-gold district that produced
around $450,000 of these metals from 1900 through
1935. In addition to copper and gold, the district
produced around 10,000 ounces of silver, mainly as
a byproduct of copper-gold mining operations at the
Lone Star and Washington and Lucile Dreyfus mines.
Although the ore from these mines averaged 1 ounce
or less per ton in silver, some prospects in the district
contain up to 42.9 ounces per ton. Currently (1975),
metal production in the Curlew district is at a standstill; however, several properties are under investigation for copper and gold.

Principal Silver Deposits

Anchor

Location: T. 39 N., R. 34 E., (section
unknown).

Development: Unknown.

Geology: Vein in dioritic rocks. Assays of 22
samples averaged 3.05 percent lead,
42.9 ozs. silver, and 0.89 oz. gold.

Ore minerals: Galena and pyrite.

Production: None.

Reference: Huntting, 1956, p. 288.

Panama

Location: Near N$\frac{1}{4}$ cor. sec. 6, T. 39 N.,
R. 34 E.

Development: Unknown.

Geology: Vein in serpentine and diorite.
Assays of 22 samples averaged 1.71
percent copper, 24.80 ozs. silver,
and 0.98 oz. gold.

Ore minerals: Galena, sphalerite, chalcopyrite, and pyrite.

Production: None.

Reference: Huntting, 1956, p. 123.

REPUBLIC DISTRICT

The Republic district of northwestern Ferry
County is the state's leading silver-producing district.
Currently (1975) 97 percent of the silver mined in
Washington is a byproduct of gold mining operations
at the Knob Hill mine near Republic. Mining began
in the district in 1897, and to date the gold mines of
the district have produced around 2 million ounces of
gold, and nearly 10 million ounces of silver. Although the Republic district is not classed as a silver
district, the nature of the deposits is discussed briefly
because of the large amount of silver produced.

Most gold-silver veins of the district occur
in a north-trending belt that is up to 1 mile wide and
6 miles long. The belt extends 4 miles north of
Republic and 2 miles south. The veins are moderately
to steeply dipping, north to northwest-trending quartz
fissure veins that average about 4 feet in thickness;
some veins are as much as 50 feet thick. Host rocks
for the veins are brecciated and altered andesitic
lava flows of the Sanpoil Volcanics (Eocene-Oligo-
cene). The most productive veins appear to be confined to the lower part of the Sanpoil Volcanics. As
well as quartz, which is the most common gangue
mineral, the veins contain lesser amounts of calcite,
adularia, laumonite, and fluorite. In addition to the
massive-banded quartz veins, some deposits consist
of brecciated quartz that has been recemented by
second generation quartz.

The ore minerals are extremely fine grained,
and occur in thin, irregular black bands and stringers

that parallel the walls of the veins. A variety of ore minerals have been reported in the veins of the district. These include free gold, electrum, chalcopyrite, tetrahedrite, stibnite, realgar, pyrargyrite, argentite, umangite, stephanite, naumannite, and native silver. The ore minerals are almost always accompanied by pyrite, which is the most plentiful sulfide. According to Full and Grantham (1968, p. 1492) high-grade ore from the Knob Hill mine contains stephanite, naumannite, native silver, and electrum. Small light-red translucent crystal suggest the presence of pyrargyrite or proustite. Most veins in the mine also contain small amounts of chalcopyrite.

The ore shoots in the district vary considerably in size; the maximum stope length is around 350 feet, whereas, the maximum pitch length is 1,300 feet. The gold and silver content of the ore shoots varies within a given vein and from mine to mine; however, the silver content of the ore almost always exceeds gold. Ore mined in 1902 to 1951 from the district averaged about 0.34 ounces per ton in gold and 2.18 ounces in silver, with a gold-silver ratio of 1 to 6.4. However, in some mines of the district gold-silver ratio was as high as 1:12.

Silver in significant amounts has been reported at only one silver property in the Republic district—the Zalla M. mine, which is 10 miles northwest of Republic. Because of the mine's proximity to the Sheridan district of Okanogan County, and because the deposit is similar to silver deposits of the Sheridan district, the Zalla M. is discussed in the Sheridan district (Okanogan County) part of this report.

OKANOGAN COUNTY

Okanogan County ranks fourth in the production of silver in Washington, having produced 860,159 ounces of the metal from 1904 through 1969. Silver

was produced in the county as early as 1889, but production figures are not available for individual counties prior to 1904. A conservative estimate of silver produced in Okanogan County from 1889 to 1904, is about 150,000 ounces.

Silver has been reported in all of the county's mining districts. It most commonly occurs at properties where gold, silver, copper, and lead are the predominant metals. A breakdown of silver occurrences according to the chief metal at each property is shown in table 22.

TABLE 22.—Distribution of silver in Okanogan County

Chief metal at property	Mines	Prospects	No. of mines or prospects
Gold	37	58	95
Silver	38	55	93
Lead	14	25	39
Copper	7	34	41
Zinc	1	4	5
Molybdenum.	1	1	2
Iron	1	1	2

Although silver occurs in most parts of Okanogan County, the most productive silver mines are in the Loomis, Conconully, Sheridan, and Nespelem districts (fig. 16). Mines that have produced over 100,000 ounces of silver are the Ruby and Kaaba-Texas in the Loomis district, and the Arlington mine in the Conconully district. Mines that have produced less than 100,000 ounces, but more than 50,000 ounces, are the First Thought in the Conconully district, Zalla M. and Sheridan in the Sheridan district, and Apache in the Nespelem district.

No major mining operations have been conducted in Okanogan County since 1950, at which

Properties

1. Mountain Sheep	39. Wind Fall
2. Ruby	40. Buck Mountain
3. Number One	41. Sherman
4. Summit	42. Silver Bell
5. Eagle	43. Silver Bluff
6. Kaaba–Texas	44. Black Huzzar
7. Chapaka	45. Eureka
8. Eureka	46. Gold Eagle
9. Rich Bar	47. Gold Cup
10. Horn Silver	48. Lulu
11. Bullfrog	49. Evening Star
12. Bellevue	50. Minnehaha
13. Kimberly	51. Okanogan Belle
14. Rainbow	52. Hardscrabble
15. Ivanhoe	53. Central
16. Leuena	54. Carl Fredrick
17. Mohawk	55. Wheeler
18. Mammoth	56. Peacock
19. Monitor	57. Shelby
20. John Arthur	58. Sheridan
21. Tough Nut	59. Zalla M.
22. Lone Star	60. American Flag
23. Key	61. Silver Bell
24. Esther	62. Eureka
25. Salmon River	63. Summit
26. Homestake	64. Mountain Boy
27. Lady of the Lake	65. Ramore
28. Leonora	66. Hanaford
29. Nevada	67. Great Divide
30. Plant–Callahan	68. Anna
31. First Thought	69. Evening
32. Last Chance	70. Andy O'Neil
33. Sonny Boy	71. Anchor
34. Fourth of July	72. Apache
35. Keystone	73. Lilman
36. Arlington	74. Little Chief
37. Woo Loo Moo Loo	75. Panama
38. Chloride	

● Mine

○ Prospect

0   6   12 Miles

Base from U.S.G.S. State of Washington
1:500,000, 1962

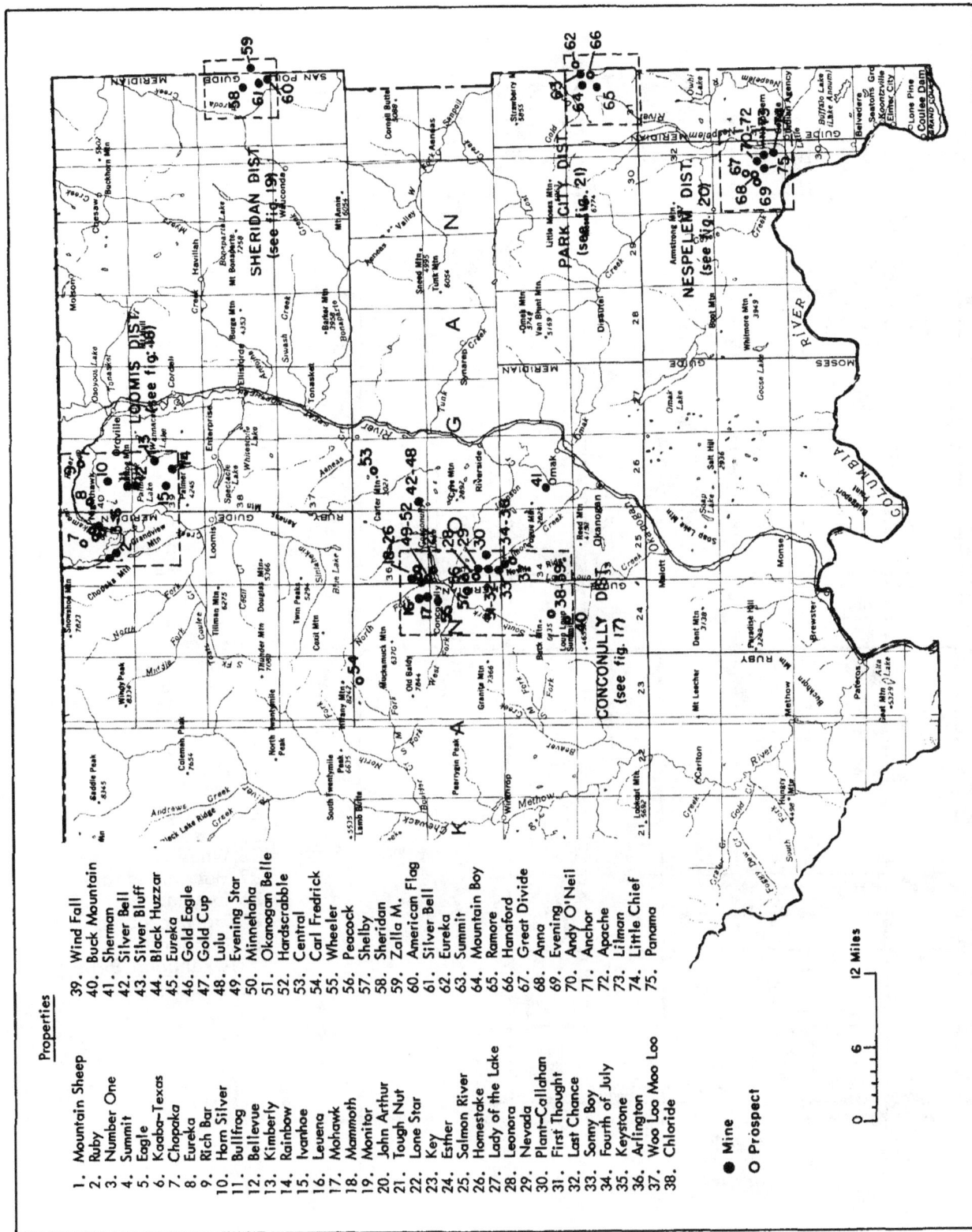

FIGURE 16.—Silver deposits of Okanogan County.

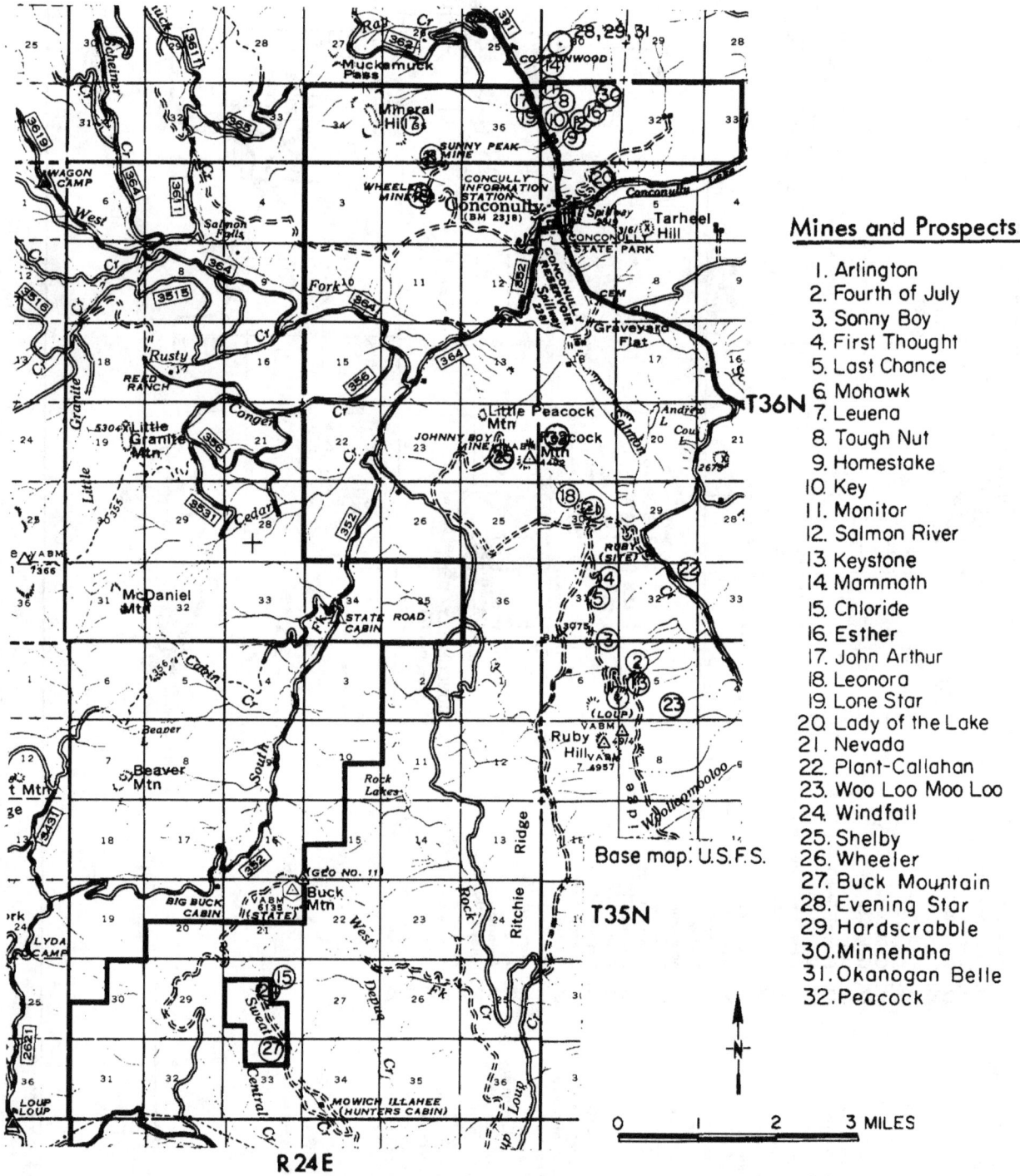

Mines and Prospects

1. Arlington
2. Fourth of July
3. Sonny Boy
4. First Thought
5. Last Chance
6. Mohawk
7. Leuena
8. Tough Nut
9. Homestake
10. Key
11. Monitor
12. Salmon River
13. Keystone
14. Mammoth
15. Chloride
16. Esther
17. John Arthur
18. Leonora
19. Lone Star
20. Lady of the Lake
21. Nevada
22. Plant-Callahan
23. Woo Loo Moo Loo
24. Windfall
25. Shelby
26. Wheeler
27. Buck Mountain
28. Evening Star
29. Hardscrabble
30. Minnehaha
31. Okanogan Belle
32. Peacock

FIGURE 17.—Index map of the Conconully district.

time the Kaaba-Texas mine closed down. However, exploration work is undertaken almost yearly at mineral deposits throughout the county. Currently (1975) several silver mines in the Loomis, Conconully, and Nespelem districts are under investigation.

CONCONULLY DISTRICT

Location

The Conconully district of central Okanogan County encompasses 525 square miles of the Okanogan Highlands physiographic province of Washington. The district falls within T. 34 N. through T. 37 N., and extends westward from the Okanogan River to the middle of R. 23 E. (fig. 17). The topography of the district consists of gently rolling, grassy range lands in the east half, and moderately to heavily timbered mountainous lands in the west half; parts of the mountainous lands are steep and rocky. Elevations near the Okanogan River are around 875 feet; however, in the west half of the the district the summits of several peaks are in excess of 6,000 feet.

Geology and Mineralization

Most silver deposits of the Conconully district occur adjacent to a contact zone between granodiorite and metamorphic rocks. Granodiorite, which is mainly in the western part of the district, is part of the Similkameen batholith (late Mesozoic). Near Conconully, where the bulk of the silver deposits are, the batholith is composed of hornblende granodiorite, biotite granodiorite, porphyritic granodiorite, quartz diorite, and quartz monzonite. The metamorphic rocks in the Conconully area are predominantly gneiss, schist, and migmatite, which are separated from the granodiorite by sharp intrusive contacts, transitional

zones of migmatitic rocks, and by the steeply dipping, northwest-trending Salmon Creek Fault. The schists and gneisses were derived from Paleozoic sediments and volcanics that were regionally metamorphosed during a Jurassic orogeny. Elsewhere in the district the predominant rocks are Triassic limestone, marble, and dolomite, and pre-Tertiary quartzite, metagraywacke, amphibolite, greenstone, and metavolcanics.

In the Conconully area the silver-bearing quartz veins vary from thin stringers less than 1 inch thick to massive quartz veins that are as much as 20 feet thick; the average quartz vein is 3 to 6 feet in thickness. The veins exhibit a wide variety of strikes ranging from N. 55° E. through N. 50° W.; the average strike is about N. 10° W. The dips of the veins are from 25° to 90°, both east and west, and average about 65°. Most veins have not been followed for much more than 500 to 600 feet along their strikes and dips.

The ore minerals of the veins, in order of decreasing abundance are galena, tetrahedrite, chalcopyrite, bornite, and sphalerite. Pyrite is almost always present in the veins; however, it appears to be more abundant in veins that contain ore minerals. The ore shoots in the veins consist of one or several of the above-mentioned minerals that occur as bands or lenses, 6 to 40 inches thick, and parallel the walls of the veins. The ore shoots are usually discontinuous, and vary in strike and pitch length from a few feet to as much as 200 feet. Sections of the veins between ore shoots usually consist of barren quartz, or quartz that contains sparsely disseminated fine-grained ore minerals.

Early reports on the mines of the Conconully area state that near-surface parts of some veins contained ore that assayed over 1,000 ounces per ton in silver, and hand-sorted ore assayed as much as 450 ounces per ton for carload lots that were shipped to smelters. However, ore that was concentrated at mills in the district averaged around 50 ounces of silver per ton. The lead content of the ore ranged from 1 to 10 percent, copper averaged 1.5 percent, and the average gold content was 0.03 ounce per ton.

Silver is the most valuable metal in the veins and occurs in argentian tetrahedrite. Bethune (1892, p. 86) reports that select samples of tetrahedrite from the First Thought mine contained up to 1,926 ounces of silver per ton. Jones (1916, p. 23) states that minerals from the Key mine contained silver as follows:

Mineral	Silver (ozs/ton)
Tetrahedrite	347.6
Galena	71.1
Sphalerite	29.5
Pyrite	26.4
Quartz	None

The vertical extent of ore minerals in the Conconully area appears to be as much as 2,500 feet on Mineral Hill, and 1,500 feet on Ruby Hill. Near the surface, oxidation is only slight; for the most part, primary ore minerals extend to the surface. Secondary enrichment does not appear to be present at any mines in the district. Near-surface parts of the veins tend to be rich in tetrahedrite, and at increased depths, galena and sphalerite predominate. At deeper depths, these minerals give way to chalcopyrite and bornite, and finally to pyrrhotite.

Production

Silver mining began in the district in 1889; and by 1893, $200,000 mainly in silver had been produced from the Arlington, First Thought, Last Chance, and Fourth of July mines. The next most

productive years were 1937 through 1939, at which time the Arlington produced $71,683 in silver and copper. The last production from the district was from 1958 through 1964, when about $12,000 in silver ore was mined at the Fourth of July. Because of incomplete production records, it is impossible to state with any accuracy the total value of metals mined in the Conconully district. A conservative estimate for the production of silver, lead, copper, and gold is around $350,000. The bulk of this production came from the silver mines that lie within 6 miles of the town of Conconully.

Major Mines

Arlington

This property, on the western slope of Ruby Hill, was the major producing metal mine of the Conconully district. Mining at the Arlington began in 1888; and by 1893, 1,000 tons of silver ore, having a net value of $25,000, had been mined. Between 1914 and 1921, several thousand tons of ore, which contained 66.6 ounces of silver per ton, returned a net profit of $31,000. In 1938 and 1939, a total of 5,700 tons of ore was mined that had a net value of $71,683. From 1888 to 1940, the mine produced $127,603 which was mainly in silver. In 1937, Arlington Mines, Inc. built a 50-ton flotation mill at the property. The mill operated in 1938 and 1939, and in 1958 it was destroyed by a fire.

The Arlington is developed by 4,500 feet of crosscuts, drifts, and shafts, as well as several stopes. The main haulage adit is 1,290 feet long and intersects the Arlington vein 972 feet from the adit's portal. The vein has been drifted upon for several hundred feet, and stoped overhead for at least 600 feet along the vein. A winze has been sunk on the vein for 100 feet beneath the main adit, and working levels are at 50 and 100 feet below the collar of the winze. About 200 feet above the main adit, an upper adit intersects the vein 420 feet from its portal, and 200 feet beneath the outcrop of the vein. This adit contains over 700 feet of drifts, and a shaft that extends to the surface. The shaft contains several working levels from which ore has been extracted. Work thus far undertaken at the mine indicates that ore extends over a vertical distance of 540 feet on the vein, and possibly deeper.

The Arlington vein is a quartz fissure vein that follows a granodiorite-gneiss contact. The vein is 1 to 6 feet thick, strikes north, and dips 70° W. Argentian tetrahedrite, chalcopyrite, galena, and minor sphalerite form bands, pods, and lenses in the vein. Parts of the vein that were mined contained ore shoots up to 2 feet thick, and as much as 300 feet long. Steeply dipping transverse faults offset parts of the vein up to 10 feet, while steeply dipping faults parallel to the vein have brecciated the ore minerals, and drawn them out into thin dark bands.

Based on past production records, the average silver content of the Arlington vein was 60 ounces per ton. Some ore contained only 10 ounces per ton in silver, while as much as 1,000 ounces were present in high-grade ore. The gold content of average ore was only 0.02 ounce, and the copper content was 1.15 percent. The richest ore shoots in the Arlington appear to have been mined during the early history of the mine; however, all parts of the vein have yet to be explored. Huntting (1956, p. 300) reports that the 100-foot level of the winze in the main adit contains two ore shoots that total 350 feet in length, average 2.5 feet thick, and contain 18 ounces per ton in silver.

Fourth of July

This property was one of the first discoveries on Ruby Hill, and as early as 1889, high-grade silver ore was shipped to a Montana smelter. So favorable were the shipments, that in 1890, a Montana syndicate acquired the mine, and undertook extensive development work. After producing $36,000 in silver, the mine shut down during the silver panic of 1893. Except for small-scale mining operations from 1958 through 1964, the mine has remained idle.

Underground workings at the Fourth of July consist of a 500-foot, double-compartment vertical shaft, and a steeply inclined 200-foot, single-compartment shaft, both of which contain several working levels. The working levels, which are up to 300 feet beneath the surface and total 850 feet in length, contain several stopes. Currently, the 500-foot shaft is caved at its collar, but the 200-foot shaft is accessible to the 100-foot level.

The Fourth of July vein is a quartz fissure vein in biotite gneiss. It averages 6 feet in thickness, strikes N. 10° W. and dips 70° to 80° E. The ore minerals consist of argentian tetrahedrite, and galena, which are almost always accompanied by pyrite. These minerals appear to be confined mainly to a 2-foot-thick section of the vein that parallels the hanging wall. Parts of the vein have been intensely sheared by faults that parallel the walls. Based on past production records, the silver content of ore shoots in the Fourth of July vein average 50 to 150 ounces per ton in silver.

First Thought

This property, on the north end of Ruby Hill, was one of the first producing mines in the Conconully district. From October 1892 through May 1893, the mine produced silver ore valued at $66,000. A 1½-mile aerial tramline carried ore from the mine to First Thought Silver Mining Co's. concentrating mill at Ruby City. Mining at the property ceased during the silver panic of 1893. Several small shipments of ore were made in the 1920's, but production was minor, and totaled only around 100 tons. Three adits and several drifts explore the vein over a vertical distance of 350 feet, and underground workings at the mine exceed 4,000 feet. In the two upper adits, drifts along the footwall and hanging wall of the vein follow it for a total strike length of about 1,000 feet. In addition to the adits, the First Thought vein has been explored by at least four shafts, which are caved at their collars.

Ore shoots at the mine contain galena, argentian tetrahedrite, sphalerite, chalcopyrite, and pyrite concentrated into lenses and bands in discontinuous lenticular masses of quartz. The lenses of quartz have general strikes of N. 10° E., dip 55° to 60° E., and occur in highly foliated feldspathic, quartzitic, and micaceous gneisses. One quartz lens is 90 feet thick at its maximum thickness, and is 600 to 700 feet in maximum breath. Individual ore shoots that were mined in the past were up to 5 feet thick, and had 75 to 100 ounces per ton in silver. Massive quartz between the ore shoots is reported to contain 6 to 8 ounces per ton in silver.

Principal Silver Deposits of Conconully District

Arlington

Location: NE¼SE¼ sec. 6, T. 34 N., R. 25 E.

Development: 4,500 feet of adits, drifts, shafts, and crosscuts. Four working levels over a vertical distance of 450 feet. Main haulage level is 1,290 feet long.

Geology: Quartz vein 1 to 6 feet thick in gneiss and granodiorite. Several ore

shoots were 2 feet thick, and up to 300 feet long. Ore averaged 60 ozs. per ton in silver, 0.02 oz. gold, and 1.15 percent copper. High-grade ore shoots assayed up to 1,000 ozs. per ton in silver.

Ore minerals: Argentian tetrahedrite, galena, chalcopyrite, and minor sphalerite.

Production: 1888-1873: $25,000 mainly in silver. 1914-1921: $31,000 mainly in silver. 1938-1939: $71,683 in silver, copper, and gold.

References: Moen, 1973, p. 10-13; Patty, 1921, p. 250-253.

Black Huzzar

Location: NE¼ sec. 31, T. 36 N., R. 26 E.

Development: Unknown.

Geology: 3½-foot quartz vein, parts of which assay 90 ozs. silver, 3.5 percent copper, and trace of gold.

Ore minerals: Minor argentite and chalcocite.

Production: None.

Reference: Huntting, 1956, p. 301.

Buck Mountain

Location: NW¼NE¼ sec. 33, T. 34 N., R. 24 E.

Development: Two 100-foot adits. Several trenches and shallow prospect pits.

Geology: Quartz vein 2 to 36 inches thick in granodiorite.

Ore minerals: Sparsely scattered grains of of tetrahedrite, pyrite, and fine-grained scheelite.

Production: Minor production in the early 1900's.

References: Huntting, 1956, p. 344; Moen, 1973, p. 39-40.

Carl Frederick (Bernhardt)

Location: Near E¼ cor. sec. 3, T. 36 N., R. 23 E.

Development: 400-foot adit with 38-foot winze. Also a 270-foot adit.

Geology: 4- to 24-inch-thick quartz vein in granodiorite. Ore shoots up to 1 foot thick contain 20 ozs. of silver per ton.

Ore minerals: Galena and pyrite.

Production: Unknown.

References: Huntting, 1956, p. 301; Patty, 1921, p. 256-257.

Central

Location: Center of sec. 10, T. 36 N., R. 26 E.

Development: 50-foot shaft and several short adits.

Geology: Sparsely metallized zone up to 300 feet wide in aplite porphyry dike that intrudes argillite and hornblende porphyry. Parts of metallized zone contains 6 to 8 ozs. per ton in silver.

Ore minerals: Galena, sphalerite, chalcopyrite, and pyrite.

Production: None.

Reference: Huntting, 1956, p. 301.

Chloride

Location: NE¼ sec. 28, T. 34 N., R. 24 E.

Development: 100-foot adit.

Geology: 4-foot quartz vein in granodiorite. Select samples from vein contained 240 to 650 ozs. silver and 1.25 to 2.0 percent copper.

Ore minerals: Argentite, chalcocite, stephanite, malachite, and pyrite.

Production: None.

Reference: Huntting, 1956, p. 302.

Esther

Location: SW¼NE¼ sec. 31, T. 36 N., R. 25 E.

Development: 74-foot adit.

Geology: 1- to 3-foot-thick quartz vein in mica schist.

Ore minerals: Galena, sphalerite, chalcopyrite, and pyrite.

Production: None.

Reference: Huntting, 1956, p. 218.

Eureka

Location: NE¼ sec. 31, T. 36 N., R. 26 E.

Development: Unknown.

Geology: 4-foot quartz vein. Select samples assayed 370 ozs. silver, 0.10 oz. gold, and 3.9 percent copper.

Ore minerals: Argentite and chalcocite.

Production: None.

Reference: Huntting, 1956, p. 302.

Evening Star

Location: Sec. 31, T. 36 N., R. 25 E.

Development: 600-foot shaft.

Geology: 4-foot quartz vein.

Ore minerals: Chalcopyrite and native silver.

Production: None.

Reference: Huntting, 1956, p. 303.

First Thought

Location: Center NE¼ sec. 31, T. 35 N.,
R. 25 E.

Development: Three adits and several shafts,
with combined underground workings
in excess of 4,000 feet. Upper adit
has around 2,000 feet of workings.

Geology: Quartz lenses up to 90 feet thick
in gneiss. Maximum breadth of lenses,
700 feet. Ore shoots contained 75 to
100 ozs. per ton in silver, massive
quartz between shoots contained 6 to
8 ozs. of silver.

Ore minerals: Argentian tetrahedrite, chalcopyrite, galena, sphalerite, and
pyrite.

Production: 1892-1893: $66,000, mainly in
silver.

References: Patty, 1921, p. 255; Jones, 1916,
p. 31-32; Bethune, 1891, p. 49-50;
Bethune, 1892, p. 79-80; Moen,
1973, p. 16-17.

Fourth of July

Location: Center NW¼ sec. 5, T. 34 N.,
R. 25 E.

Development: 500-foot, double-compartment
shaft, and a 200-foot, single-compartment shaft with at least 850 feet of
workings on five levels with stopes on
most levels.

Geology: 6-foot quartz vein in biotite gneiss.
Ore shoots up to 2 feet thick contained
50 to 150 ozs. per ton in silver. Ore
mined from 1958 through 1964 averaged 60 ozs. per ton in silver.

Ore minerals: Argentian tetrahedrite, galena,
and pyrite.

Production: 1890-1893, 1958-1964: total
production around $50,000 mainly in
silver.

References: Moen, 1973, p. 13-14; Bethune,
1891, 49-50; Bethune, 1892, p.79-80.

Gold Cup

Location: NE¼ sec. 31, T. 36 N., R. 26 E.

Development: Unknown.

Geology: Quartz vein, parts of which assay
80 ozs. per ton in silver.

Ore minerals: Unknown.

Production: Unknown.

Reference: Huntting, 1956, p. 303.

Gold Eagle

Location: NE¼ sec. 31, T. 36 N., R. 26 E.

Development: Unknown.

Geology: Quartz vein, parts of which assay
100 ozs. per ton in silver.

Ore minerals: Unknown.

Production: None.

Reference: Huntting, 1956, p. 303.

Hardscrabble

Location: Sec. 30, T. 36 N., R. 25 E.

Development: 30-foot drift.

Geology: 15-inch vein with 11-inch metallized zone. Assays of 178,335 and
605 ozs. per ton in silver reported.

Ore minerals: Stephanite and pyrite.

Production: None.

Reference: Huntting, 1956, p. 304.

Homestake

Location: E½SW¼ sec. 31, T. 36 N.,
R. 25 E.

Development: 29-foot shaft, and a 175-foot
adit, with a 42-foot crosscut.

Geology: 11-foot-thick quartz vein in quartz-
mica schist. Ore shoots contained 18
to 56 ozs. per ton in silver, and 32 to
36 percent lead.

Ore minerals: Galena and pyrite.

Production: 400 tons prior to 1902, of which
100 tons netted $1,500.

References: Huntting 1956, p. 220; Bethune,
1891, p. 56.

John Arthur

Location: SE¼NE¼ sec. 36, T. 36 N.,
R. 24 E.

Development: 70-foot inclined shaft.

Geology: 10-foot-thick quartz vein, 2 to
2½ feet of which is metallized. Parts
of vein assay 30 to 116 ozs. per ton
in silver, and traces to 5 ozs. gold.

Ore minerals: Argentite, native silver, and
pyrite.

Production: Unknown.

Reference: Huntting, 1956, p. 305.

Key

Location: SW¼NW¼ sec. 31, T. 36 N.,
R. 25 E.

Development: 300-foot adit, 105-foot drift,
and 80-foot shaft.

Geology: Sparsely metallized quartz vein 3
to 10 feet thick in mica schist and
migmatite. Vein contained small,
scattered ore shoots, with up to several hundred ozs. per ton in silver.

Ore minerals: Galena, sphalerite, and chalcopyrite.

Production: 1,500 tons prior to 1914.

References: Huntting, 1956, p. 304; Jones,
1916, p. 25; Gage, 1941, p. 207.

Keystone

Location: S½NW¼ sec. 5, T. 34 N.,
R. 25 E.

Development: 150-foot shaft.

Geology: 8-foot-thick quartz vein in biotite
gneiss. Parts of vein contain up to
50 ozs. per ton in silver.

Ore minerals: Tetrahedrite and galena.

Production: Unknown.

References: Huntting, 1956, p. 305; Bethune,
1891, p. 52-53.

Lady of the Lake

Location: Center NE¼ sec. 6, T. 35 N.,
R. 25 E.

Development: 150-foot adit with 50 feet of
drifts. Several open cuts.

Geology: Shear zones up to 20 feet wide in
mica schist. Sparsely metallized but
select samples contained 70 to 125 ozs.
silver and 11 to 15 percent lead.
Average sample contains less than 1
percent each of copper, lead, and
zinc.

Ore minerals: Galena, pyrite, chalcopyrite,
sphalerite, molybdenite, and scheelite.

Production: None.

Reference: Huntting, 1956, p. 305.

Last Chance

Location: Center sec. 31, T. 35 N., R. 25 E.

Development: 300-foot shaft with working
levels at 100, 200, and 300 feet.
300-foot level contains 800 foot crosscut; 650 foot adit with stopes.

Geology: 12-foot-thick quartz vein in schist
and granodiorite. Ore shoots 2 to 4
feet thick, and up to 200 feet long,
averaged 30 ozs. of silver per ton,
17 percent lead, and 4 percent copper.

Ore minerals: Argentian tetrahedrite, galena,
chalcopyrite, sphalerite and pyrite.

Production: Minor production in 1920, 1921,
and 1924.

References: Patty, 1921, p. 254-255; Moen,
1973, p. 17-18.

Leonora

Location: Sec. 30, T. 35 N., R. 25 E.

Development: 80-foot shaft.

Geology: 5-foot-thick quartz vein with a
4-inch paystreak that assays up to
350 ozs. silver and 17 to 25 percent
lead.

Ore minerals: Galena, stephanite, and pyrite.

Production: None.

Reference: Huntting, 1956, p. 220.

Leuena

Location: SE¼SW¼ sec. 35, T. 36 N.,
R. 24 E.

Development: 50- and 75-foot shafts, 60-
foot adit, and 70-foot crosscut.

Geology: 7-foot-thick quartz vein in granodiorite. Ore shoots contained 200 to
800 ozs. per ton in silver.

Ore minerals: Tetrahedrite, stephanite, and
argentite.

Production: Several carloads prior to 1890.

References: Huntting, 1956, p. 305; Bethune,
1891, p. 59; Jones, 1916, p. 28-29.

Lone Star

Location: NE¼ sec. 36, T. 36 N., R. 24 E.

Development: 300-foot inclined shaft and
2 adits with total of 2,000 feet of
workings.

Geology: 1½- to 20-foot-thick quartz vein
in granodiorite. Select samples contained up to 65 percent lead, 140 ozs.
silver, and 1.6 oz. gold. Samples of
1½- to 10-foot-thick vein assayed
1.85 to 7.35 ozs. silver, 0.1 to 1.1
percent zinc, and 0.1 to 0.6 percent
lead.

Ore minerals: Sparse galena, sphalerite,
chalcopyrite, pyrite, and stephanite.

Production: 40 tons in 1913.

References: Huntting, 1956, p. 221; Moen,
1973, p. 24-26.

Lulu

Location: NE¼ sec. 31, T. 36 N., R. 26 E.

Development: Unknown.

Geology: 4-foot thick quartz vein, parts of
which contain up to 320 ozs. per ton
in silver.

Ore minerals: Argentite and chalcocite.

Production: None.

Reference: Huntting, 1956, p. 306.

Mammoth

Location: SW¼ sec. 30, T. 36 N., R. 25 E.

Development: Two 30-foot crosscuts.

Geology: 5- and 6-foot quartz veins in mica schist. High-grade ore contained 87 to 330 ozs. per ton in silver, 1 to 3 ozs. of gold, and 7 percent copper.

Ore minerals: Galena, chalcopyrite and pyrite.

Production: Small shipment in 1889 netted $250 per ton.

Reference: Bethune, 1892, p. 95.

Minnehaha

Location: Sec. 31, T. 36 N., R. 25 E.

Development: 200-foot crosscut.

Geology: 17-foot-thick quartz vein. Select samples assay $12\frac{1}{2}$ to 14 percent lead and 98 to 113 ozs. silver.

Ore minerals: Galena, tetrahedrite, and pyrite.

Production: None.

Reference: Huntting, 1956, p. 306.

Mohawk (Chief Sunshine)

Location: SW$\frac{1}{4}$SE$\frac{1}{4}$ sec. 35, T. 36 N., R. 24 E.

Development: 500- and 150-foot adits.

Geology: $1\frac{1}{2}$- to 3-foot-thick quartz vein in porphyritic granodiorite. Ore shoots up to 1 foot wide contained up to 60 ozs. per ton in silver, 13 percent lead, and 3 percent copper.

Ore minerals: Galena, chalcopyrite, tetrahedrite, and pyrite.

Production: 1890-1893, 1951-1954, 1961 and 1967: total production probably less than 1,000 tons.

References: Moen, 1973, p. 24; Bethune, 1891, p. 60

Monitor

Location: NW$\frac{1}{4}$NW$\frac{1}{4}$ sec. 31, T. 36 N., R. 25 E.

Development: 200-foot adit and 50-foot shaft.

Geology: 2-foot-thick quartz vein in mica schist. Ore shoots contained up to 35 ozs. per ton in silver, and 10 percent lead.

Ore minerals: Galena, sphalerite, and chalcopyrite.

Production: Minor production prior to 1900. Shipment in 1889 netted $250 per ton in silver.

References: Huntting, 1956, p. 309; Jones, 1916, p. 26; Bethune, 1891, p. 56-58.

Nevada

Location: SW$\frac{1}{4}$NE$\frac{1}{4}$ sec. 30, T. 36 N., R. 25 E.

Development: 1,000-foot adit with drifts and crosscuts and 4 shafts up to 220 feet deep.

Geology: 3- to 5-foot-thick quartz vein in schist and gneiss. Ore minerals occur in pod and bands and ore shoots average 4.5 ozs. silver, 3.5 to 8.6 percent lead, 3.15 percent zinc, and 0.2 to 0.7 percent copper.

Ore minerals: Galena, tetrahedrite, pyrite, chalcopyrite, sphalerite, and stromeyerite.

Production: Minor production 1954-1957.

References: Huntting, 1956, p. 307; Moen, 1973, p. 18-20.

Okanogan Belle

Location: Sec. 30, T. 36 N., R. 25 E.
Development: 150-foot crosscut adit.
Geology: 13-foot-thick quartz vein. Select samples from vein and dump assay 92 to 95 ozs. silver and 10 percent lead.
Ore minerals: Minor galena, argentite, and pyrite.
Production: None.
Reference: Huntting, 1956, p. 307.

Peacock

Location: Sec. 19, T. 35 N., R. 25 E.
Development: 170-foot crosscut adit and shaft.
Geology: Quartz veins along contact between gneiss and granodiorite. Vein assays 10.4 to 19.2 ozs. silver, 0.5 to 0.8 percent copper, and 2.2 to 18 percent lead.
Ore minerals: Sparse galena, tetrahedrite, chalcopyrite, sphalerite, and pyrite.
Production: Minor production in the 1920's.
Reference: Huntting, 1956, p. 308.

Plant Callahan

Location: NE$\frac{1}{4}$NE$\frac{1}{4}$ sec. 32, T. 35 N., R. 25 E.
Development: 65- and 175-foot adits.
Geology: Narrow quartz veins in biotite gneiss. Select high-grade samples contained up to 1,000 ozs. per ton in silver.
Ore minerals: Tetrahedrite and pyrite.
Production: Minor prior to 1900.
Reference: Huntting, 1956, p. 308.

Salmon River

Location: NE$\frac{1}{4}$SW$\frac{1}{4}$ sec. 31, T. 36 N., R. 25 E.
Development: 150-foot crosscut adit with 500 feet of drifts. Also 20 and 30-foot adits and 20-foot crosscut.
Geology: Four quartz veins 1 to 5 feet thick in mica schist. Parts of some veins contained up to 4 ozs. per ton in gold, 83 ozs. per ton in silver, and 20 percent lead.
Ore minerals: Argentite, tetrahedrite, galena, sphalerite, and chalcopyrite.
Production: Unknown.
References: Huntting, 1956, p. 309; Jones, 1916, p. 26; Bethune, 1891, p. 56-58.

Shelby

Location: "West side of Peacock Hill." Possibly same as Johnny Boy mine in center sec. 24, T. 35 N., R. 24 E.
Development: 75-foot inclined shaft.
Geology: 6-foot-thick quartz vein in granodiorite. Select samples of vein assayed 29 to 140 ozs. silver and 0.5 to 5 percent lead.
Ore minerals: Galena, arsenopyrite, and pyrite.
Production: None.
Reference: Huntting, 1956, p. 309.

Sherman

Location: SW$\frac{1}{4}$SE$\frac{1}{4}$ sec. 20, T. 34 N., R. 26 E.
Development: 220-foot shaft with over 400 feet of drifts and crosscuts. Also

100- and 40-foot incline shafts.

Geology: 2- to 11-foot-wide metallized shear zones in granite and rhyolite. Parts of the shear zones contained 3.9 to 31 ozs. silver, traces to 0.08 oz. gold, 1.7 to 7.7 percent lead, and 0.67 to 8.1 percent zinc.

Ore minerals: Galena, sphalerite, chalcopyrite, arsenopyrite, and pyrite.

Production: Minor production in 1958.

References: Huntting, 1956, p. 223; Moen, 1973, p. 38-39.

Silver Belle

Location: NE¼ sec. 31, T. 36 N., R. 26 E.

Geology: Quartz vein, 2½ feet thick.

Ore minerals: Chalcocite, argentite, and stromeyerite.

Production: None.

References: Huntting, 1956, p. 310; Moen, 1973, p. 34.

Silver Bluff

Location: NE¼ sec. 31, T. 36 N., R. 26 E.

Development: 65-foot inclined shaft.

Geology: 2-foot-thick quartz vein, parts of which contained up to 110 ozs. per ton in silver, and 3.5 percent copper.

Ore minerals: Argentite and chalcocite.

Production: $80,000 to end of 1923.

References: Huntting, 1956, p. 310; Bethune, 1892, p. 112.

Sonny Boy

Location: S½SE¼ sec. 31, T. 35 N., R. 25 E.

Development: 600-foot main adit with several

small stopes, crosscuts, and raises. Surface stope above adit.

Geology: ½- to 6-foot-thick quartz veins in granodiorite. High-grade ore shoots contained up to several hundred ozs. per ton in silver. Existing veins are only sparsely metallized.

Ore minerals: Tetrahedrite, galena, chalcopyrite, sphalerite, and pyrite.

Production: Minor production in 1937 and and 1939.

References: Moen, 1973, p. 14-16.

Sunrise

Location: "On Mineral Hill."

Development: 75-foot adit.

Geology: 4-foot-thick quartz vein with a 14-inch paystreak. High-grade samples assayed 379 ozs. silver per ton.

Ore minerals: Argentite, stephanite, and tetrahedrite.

Production: Unknown.

Reference: Huntting, 1956, p. 311.

Tough Nut

Location: Center NW¼ sec. 31, T. 36 N., R. 25 E.

Development: 50-foot inclined shaft, and a 250-foot adit, with a 40-foot winze.

Geology: 3- to 10-foot-thick quartz vein in quartz-mica schist. Ore shoots contained 47 to 86 ozs. per ton in silver, and 32 to 43 percent lead.

Ore minerals: Galena, sphalerite, and chalcopyrite.

Production: $9,000 prior to 1901.

References: Bethune, 1891, p. 55-56; Jones, 1916, p. 25; Gage, 1941, p. 209.

Wheeler (Mineral Hill)

Location: Center sec. 2, T. 35 N., R. 24 E.

Development: 3,000 feet of drifts, crosscuts, and shafts. Main adit consists of a 1,200-foot drift with a 1,400-foot crosscut. A 135-foot shaft from surface to main adit. Also several short adits.

Geology: Quartz veins from 1 to 36 inches thick in granodiorite. Columbia vein contained ore that averaged 34 ozs. per ton in silver, 0.22 oz. gold, and 10 percent lead. A 100-foot-thick molybdenite-bearing dike reported on property.

Ore minerals: Galena, chalcopyrite, sphalerite, tetrahedrite, and pyrite occur as disseminated grains in the quartz veins.

Production: Minor production in 1938 and 1939.

References: Moen, 1973, p. 20-24.

Windfall

Location: Sec. 28, T. 34 N., R. 24 E.

Development: 60-foot shaft.

Geology: 3-foot-thick quartz vein in granodiorite. Select samples of vein assayed 118 to 325 ozs. silver per ton.

Ore minerals: Stephanite and cerargyrite.

Production: None.

Reference: Huntting, 1956, p. 311.

Woo Loo Moo Loo

Location: NE$\frac{1}{4}$NE$\frac{1}{4}$ sec. 5, T. 34 N. R. 25 E.

Development: 150-foot shaft.

Geology: 8-foot-thick quartz vein in biotite-hornblende gneiss.

Ore minerals: Galena and tetrahedrite.

Production: None.

References: Huntting, 1956, p. 305; Moen, 1973, p. 27.

LOOMIS DISTRICT

Location

For the purpose of this report the Loomis district includes the Nighthawk, Oroville, Palmer Mountain, and Wannacut Lake districts of north-central Okanogan County. These combined districts extend from the Okanogan River westward to the west border of R. 23 E., and from the international border southward to the north border of T. 37 N. For the most part the district is mountainous, with several peaks in the west half of the district exceeding 6,000 feet in elevation. In the east half, the elevations of the main peaks are 3,000 to 5,000 feet. Near the central part of the district, elevations along the Sinlahekin Valley range from 1,200 to 1,700 feet; while along the Okanogan River, elevations are around 900 feet. Deposits of gold, lead, and silver predominate in the Loomis district, and several thousand ounces of silver have been produced from the mines of the area.

Geology and Mineralization

Most silver mines and prospects of the Loomis district are of the quartz fissure vein type. The veins, which are believed to be middle Cretaceous to early Tertiary age, occur in Jurassic or Cretaceous granodiorite, as well as in argillite, limestone, conglomerate, quartzite, and greenstone of Permian or Triassic age. The veins range from stringers less than 1 inch to massive quartz veins 20 feet thick. They display a wide variety of strikes and dips, and have a tendency to pinch and swell. One or several of the following ore minerals are present in the veins: pyrite, arseno-pyrite, chalcopyrite, galena, sphalerite, scheelite, argentite, proustite, pyrargyrite, native silver, and free gold. In some veins the ore minerals are segregated into irregular bands that parallel the walls of the vein, while in other veins the ore minerals occur in pods or lenses. Most deposits exhibit little in the way of secondary enrichment. However, at the Ivanhoe mine, near-surface parts of the vein contained abundant silver oxides and native silver. The ore in most veins has not been mined for much more than 100 feet in stope and pitch lengths, which indicates that the majority of the deposits are small. However, at the Ruby the vein contained ore for as much as 800 feet along the strike of the vein, and for as much as 770 feet beneath the outcrop of the vein. At the Kaaba-Texas mine the vein has been mined for up to 800 feet along its strike, and to a depth of 360 feet beneath its outcrop. The maximum vertical extent of the ore minerals in the veins of the district has yet to be determined.

Some veins were extremely rich in silver, but the average silver content of the veins was only about 5 ounces per ton. Hand-sorted, near-surface ore from the Ivanhoe contained 572 ounces per ton in silver, whereas high-grade gold ore from the Rainbow mine assayed up to 200 ounces per ton in silver. The Ruby, which is a true silver mine, produced several thousand tons of ore that averaged 10 ounces per ton in silver; however, some parts of the vein contained several hundred ounces per ton in silver. The Kaaba-Texas, which was the major silver-producing mine in the district, produced 74,485 tons of lead-copper-zinc ore that averaged only 1.825 ounces per ton in silver.

Production

The major silver-producing mines were the Ruby, which produced around 100,000 ounces of silver from 1915 to 1926, and the Kaaba-Texas, which produced about 200,000 ounces from 1915 to 1951. At the Ruby, silver was the predominant metal, whereas at the Kaaba-Texas, lead, silver, zinc, and copper were the predominant metals. In addition to the Ruby and Kaaba-Texas mines, minor silver was produced during the early 1900's at the Horn Silver, Bullfrog, Golden Zone, Mountain Sheep, Number One, Bellevue, Ivanhoe, and Rainbow mines. Mining in the district ceased in 1951 with the closure of the Kaaba-Texas mine; however, minor exploration has been undertaken at several properties in the district over the past 20 years.

Major Mines

Ivanhoe mine

This mine near the summit of Palmer Mountain was for a short time a producer of some of the richest silver ore mined in Washington. Hand-sorted, near-surface ores mined from 1888 to 1897 contained up to 1,000 ounces per ton in silver, as well as several ounces of gold. Around 1,000 tons of ore was shipped to smelters in Washington and Montana; several shipments averaged 392 ounces per ton in silver and 1.20 ounces in gold.

The Ivanhoe vein was first developed by several shallow shafts, one of which was later sunk to a depth of 500 feet on the vein. Inasmuch as the richest

Mines and Prospects

1. Horn Silver
2. Ivanhoe
3. Mountain Sheep
4. Rainbow
5. Ruby
6. Kimberly
7. Kaaba–Texas
8. Bellevue
9. Bullfrog
10. Chopaka
11. Eureka
12. Eagle
13. Number One
14. Rich Bar
15. Summit

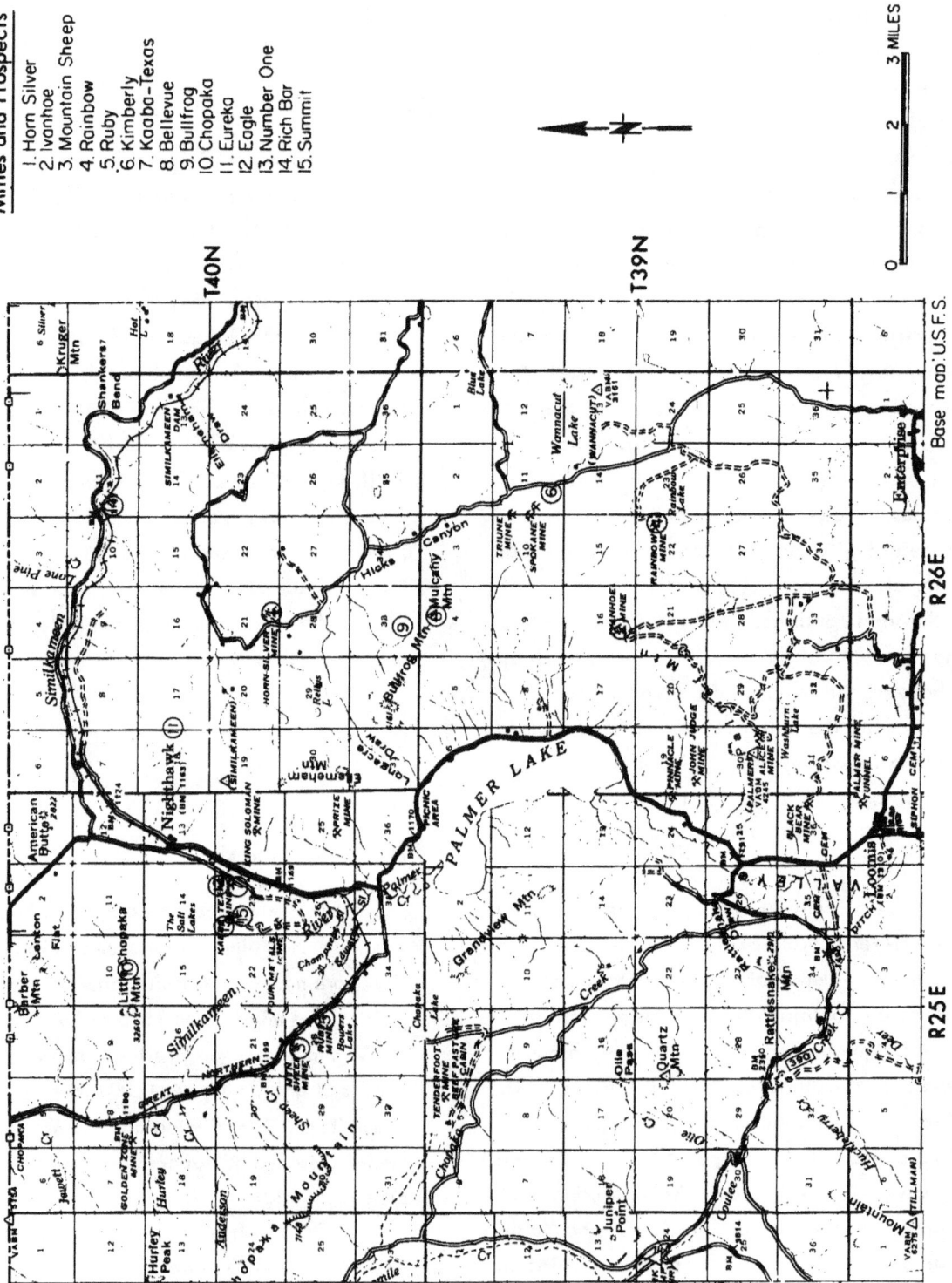

FIGURE 18.—Index map of the Loomis district.

silver ore proved to be near the surface, the ore was mined by stripping the hanging wall of the vein over an area of 70 by 120 feet, and to a maximum depth of around 30 feet. Around 1914, a 4,400-foot crosscut was driven from the west slope of Palmer Mountain to intersect the Ivanhoe vein at a depth of 1,560 feet. Although at least seven quartz veins were encountered, the average silver content of the veins proved to be less than 1 ounce per ton. Mining activity apparently ceased at the Ivanhoe around 1915. Currently (1975), only the 4,400-foot crosscut is accessible; other workings on the property are caved or full of water.

The Ivanhoe vein is a $3\frac{1}{2}$- to $4\frac{1}{2}$-foot-thick quartz fissure vein, which has for host rocks Permian carboniferous phyllite and slaty argillite. Near the surface the vein is greatly fractured and contorted. It has a general north-northwest strike, and dips 60° W. At one time the vein was stripped of its overburden for around 1,200 feet; however, very little of the vein is presently visible. The ore minerals consist of pyrite, tetrahedrite, galena, stephanite, cerargyrite, native silver, and free gold. Ore that was mined occurred in a $2\frac{1}{2}$-foot-wide ore shoot that contained dark streaks and blotches of very fine-grained ore minerals. Currently (1975), only a small part of the vein is visible at the collar of the inclined shaft on the Ivanhoe claim. Select samples from this part of the vein assayed 91.70 ounces per ton in silver. Because the 500-foot inclined shaft is caved near its collar and full of water, the silver content of other parts of the vein remains unknown. The silver content of seven quartz veins exposed in the 4,400-foot crosscut ranges from 0.34 to 1.82 ounces per ton; the gold content averages less than 0.02 ounce per ton.

Ruby Mine

This mine, which is 3 miles southwest of Nighthawk, was an important silver mine of the Loomis area,

having produced around 100,000 ounces of silver from 1915 to 1922. High-grade ore contained as much as several hundred ounces per ton in silver, but mill-run ore averaged only 10 ounces per ton in silver. In 1920, Pyrargyrite Mining Company built a 75-ton flotation mill at the mine, and from 1920 to 1923, the company produced silver concentrates. Prior to 1920, only crude ore was shipped to smelters. Except for small-scale development work, the mine is currently inactive.

The mine, which is $2\frac{1}{2}$ miles northwest of Palmer Lake, is developed by at least 5,000 feet of underground workings, and several stopes. The main adit intersects the Ruby vein 950 feet from the adit's portal, and 550 feet beneath the outcrop of the vein. At the adit level the vein has been drifted upon for 600 feet, and near the face of the northwest drift a winze was sunk 210 feet on the vein. Raises follow the vein upward from the adit level to the surface, and connect with four intermediate levels.

The Ruby vein is a quartz fissure vein in a shear zone in granodiorite of Cretaceous age. The vein averages 3 feet in thickness, strikes N. 45° W., and dips 42° SW. The ore minerals, which occur as sparsely scattered grains in the vein, consist of pyrite, arsenopyrite, chalcopyrite, galena, sphalerite, pyrargyrite, proustite, and argentite. The vein contains at least two ore shoots, parts of which have been mined; all parts of the vein have yet to be explored.

Principal Silver Deposits of Loomis District

Bellevue

Location: SE$\frac{1}{4}$NW$\frac{1}{4}$ sec. 4, T. 39 N., R. 26 E.

Development: Shallow shaft and several open cuts.

Geology: 1- to 3-foot-thick quartz vein in
slate. Hand-sorted ore contained
60 ozs. silver and 2 ozs. gold per ton.

Ore minerals: Chalcopyrite, pyrargyrite,
native silver, free gold, pyrite, and
arsenopyrite.

Production: Minor.

References: Huntting, 1956, p. 136; Rinehart and Fox, 1972, p. 78-79.

Bullfrog

Location: S½SW¼ sec. 33, T. 40 N.,
R. 26 E.

Development: 140- and 160-foot shafts and
1 adit.

Geology: 7-foot-thick quartz vein in quartzite and schist. Vein traceable for
3,000 feet and assays up to 10 ozs.
silver and 0.5 oz. gold.

Ore minerals: Pyrite and fine-grained black
sulfides.

Production: Minor production in early 1900's.

Reference: Huntting, 1956, p. 63.

Chopaka

Location: S½ sec. 10, T. 40 N., R. 25 E.

Development: 30-foot adit.

Geology: 10-foot-thick quartz vein in granodiorite. Select samples of vein
assayed 0.80 to 1.10 ozs. gold, 75
to 125 ozs. silver, and 1.5 to 1.9
percent copper.

Ore minerals: Chalcopyrite and pyrite.

Production: None.

Reference: Huntting, 1956, p. 38.

Eagle

Location: NE¼ sec. 23, T. 40 N., R. 25 E.

Development: 110-foot adit.

Geology: 12-foot-thick quartz vein. Select
samples from vein assayed 18 to 310
ozs. silver, and 10 to 12 percent lead.

Ore minerals: Galena and pyrite.

Production: None.

Reference: Huntting, 1956, p. 218.

Eureka

Location: Secs. 17 and 18, T. 40 N., R. 26 E.

Development: 7 shafts with a total of 360
feet of workings.

Geology: 3- to 7-foot-thick quartz veins in
granodiorite. Average of eight assays
of the veins showed 17.71 ozs. silver
and 7.15 percent lead.

Ore minerals: Galena and pyrite.

Production: None.

Reference: Huntting, 1956, p. 218.

Horn Silver

Location: S½ sec. 21, T. 40 N., R. 26 E.

Development: 750-foot adit and a 100-foot
shaft.

Geology: Quartz veins up to 4 feet thick in
argillite and granodiorite. Ore shoots
averaged around 110 ounces per ton
in silver.

Ore minerals: Stephanite, cerargyrite,
proustite, galena, chalcopyrite,
sphalerite, and pyrite.

Production: Several carloads prior to 1910.

References: Umpleby, 1911b, p. 100;
Huntting, 1956, p. 304.

Ivanhoe

Location: SW¼ sec. 16, T. 39 N., R. 26 E.
Development: 4,400-foot crosscut adit,
500-foot inclined shaft, and a 70-foot
vertical shaft.
Geology: 3½- to 4½-foot-thick quartz vein
in calcareous argillite. Near the
surface the vein contained a 2½-foot-
thick ore shoot that carried from 36 to
572 ounces of silver per ton. Select
samples of the vein assays up to 50
ounces per ton in silver.
Ore minerals: Pyrargyrite, cerargyrite,
stephanite, native silver, free gold,
galena, and pyrite.
Production: 1888 to 1897, around 1,000 tons
of hand-sorted, high-grade ore.
References: Bethune 1891, p. 63-64;
Bethune 1892, p. 99.

Kaaba-Texas

Location: Center NE¼ sec. 23, T. 40 N.,
R. 25 E.
Development: 300-foot inclined shaft, with
1,100 feet of drifts on four levels.
Geology: 6- to 12-foot-thick quartz vein
along lamprophyre-granodiorite contact. Ore averaged 2.25 ozs. per
ton in silver, 1.2 percent lead, ½ percent zinc, and 0.1 percent copper.
Ore minerals occur as bands along
footwall, and as lenses and pods along
hanging wall.

Ore minerals: Galena, chalcopyrite, sphalerite, molybdenite, scheelite, and
pyrite.
Production: 1918, 1943-1951: around 205,000
ozs. of silver. 1943-1946: 135,973
ozs. silver, 1,357,185 lbs. lead,
506,050 lbs. zinc, 99,410 lbs. copper.
References: Patty, 1921, p. 228-231; Gage,
1941, p. 216-219; Culver and Broughton, 1945, p. 34-35.

Kimberly

Location: Near center SW¼ sec. 11, T. 39 N.,
R. 26 E.
Development: 140-foot inclined shaft, with
drifts on the 60-, 80-, and 100-foot
levels.
Geology: En echelon quartz lenses along
contact between diorite and slaty
argillite. Ore shoots, 1 to 5 feet
wide and 40 to 50 feet long, containing 60 to 70 ozs. per ton in silver,
10 to 20 percent lead, and up to 1 oz.
of gold.
Ore minerals: Galena, sphalerite, chalcopyrite, and pyrite.
Production: Amount not known.
References: Umpleby, 1911b, p. 96-97.

Mountain Sheep

Location: NE¼NW¼ sec. 28, T. 40 N.,
R. 25 E.
Development: Three adits totaling 2,000 feet.
Geology: Quartz vein up to 6 feet thick in
granodiorite. Select ore samples
assayed up to 100 ozs. per ton in silver.

Ore minerals: Unknown.

Production: Few carloads prior to 1911.

References: Huntting, 1956, p. 307.

Number One

Location: NW$\frac{1}{4}$NW$\frac{1}{4}$ sec. 23, T. 40 N.,
R. 25 E.

Development: 200-foot adit and several
shafts, one of which connects with
the adit.

Geology: 3- to 12-foot-thick quartz vein
along fault between granodiorite, and
slate and conglomerate. Parts of vein
assays up to 50 ozs. silver and 20 percent lead.

Ore minerals: Galena.

Production: None.

Reference: Huntting, 1956, p. 221.

Rainbow

Location: Center NE$\frac{1}{4}$ sec. 22, T. 39 N.,
R. 26 E.

Development: Three adits, with a total of
850 feet of workings.

Geology: Quartz lenses in limestone, schist,
and quartzite. Ore shoots contained
up to 200 ozs. per ton in silver, two
percent copper, and up to 3$\frac{1}{2}$ percent
lead.

Ore minerals: Chalcopyrite, galena, native
gold, malachite, arsenopyrite, and
pyrite.

Production: Has produced, but amount unknown.

References: Rinehart and Fox, 1972, p.100-
101; Huntting, 1956, p. 147.

Rich Bar

Location: SW$\frac{1}{4}$NW$\frac{1}{4}$ sec. 11, T. 40 N.,
R. 26 E.

Development: 2 shafts, deepest one 150 feet,
with drifts at 50- and 150-foot levels.

Geology: Quartz veins up to 6 feet thick in
quartzitic slate.

Ore minerals: Chalcopyrite, sphalerite,
galena, stephanite, argentite, and
pyrite.

Production: None.

Reference: Huntting, 1956, p. 70.

Ruby

Location: NE$\frac{1}{4}$SE$\frac{1}{4}$ sec. 28, T. 40 N.,
R. 25 E.

Development: 950-foot adit, with 5,000 feet
of drifts and raises.

Geology: Metallized shear zone in granodiorite averages around 3 feet in thickness.
Gangue consists of quartz, calcite,
and gouge. High-grade ore averaged
several hundred ounces per ton in silver;
mill-run ore averaged around 10 ozs.
per ton.

Ore minerals: Pyrargyrite, proustite, argentite, galena, chalcopyrite, sphalerite,
native gold, arsenopyrite, and pyrite.

Production: 1915-1923: around 100,000 ozs.
of silver.

References: Huntting, 1956, p. 309; Patty,
1921, p. 237-240; Umpleby, 1911b,
p. 94-95; Gage, 1941, p. 224-225.

Summit

Location: NE$\frac{1}{4}$NW$\frac{1}{4}$ sec. 23, T. 40 N.,
R. 25 E.

Development: 70-foot inclined shaft.

Geology: 3½- to 4-foot-thick quartz vein in granodiorite. Vein assays 5 to 7 ozs. per ton in silver.

Ore minerals: Galena, sphalerite, chalcopyrite, molybdenite, and pyrite.

Production: None.

References: Huntting, 1956, p. 310; Rinehart and Fox, 1972, p. 104-105.

SHERIDAN DISTRICT

Location

The Sheridan district is about 10 miles northwest of Republic, in northeastern Okanogan County. Most silver occurrences are immediately west of the Ferry-Okanogan county line, and in sections 23, 24, 25, and 26, T. 38 N., R. 31 E. The Zalla M. mine, which produced rich silver ore, is 0.2 miles east of the district's eastern border, but because of its proximity and similarity to silver deposits of the Sheridan district, it is included in this discussion.

Geology and Mineralization

The silver deposits of the Sheridan district occur in quartz fissure veins, and in narrow shear zones in phonolite of the Klondike Mountain Formation of Oligocene and Miocene(?) age. Although granitic rocks do not crop out at the silver mines, they form extensive outcrops ½ to 1½ miles east of the mines. High-angle north-trending faults separate the granitic rocks (Cretaceous or older) from the Klondike Mountain Formation.

At the Sheridan and American Flag mines, unidentified silver minerals occur in 1- to 8-foot-wide

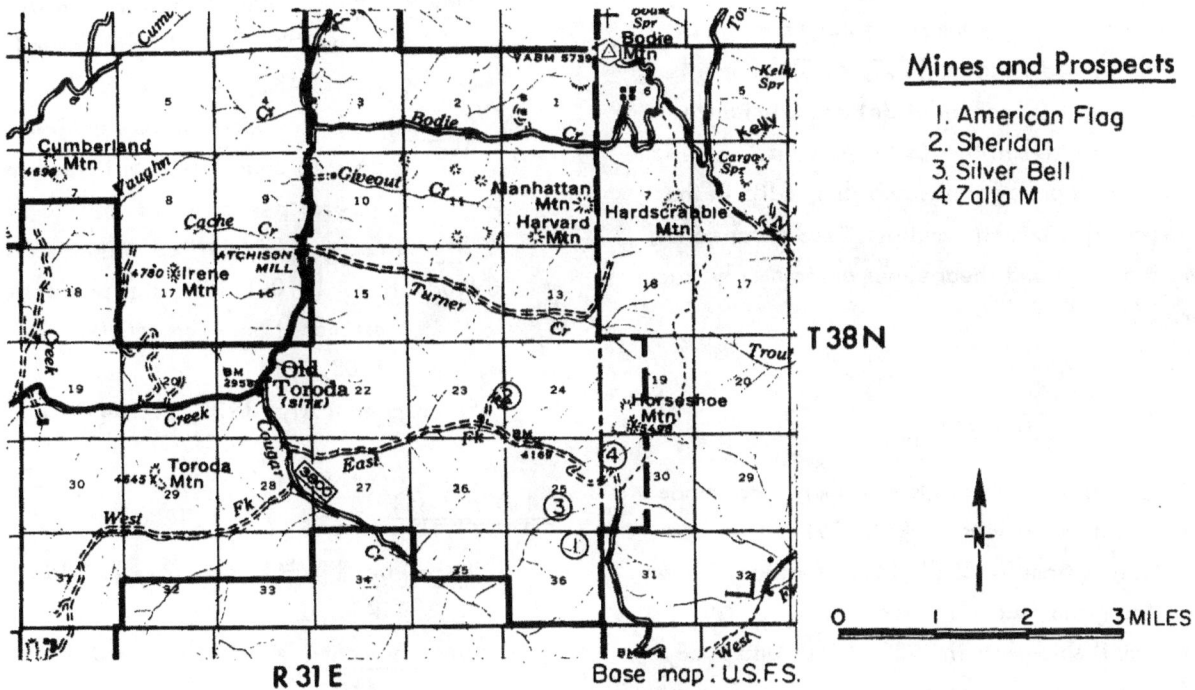

Mines and Prospects

1. American Flag
2. Sheridan
3. Silver Bell
4. Zalla M

FIGURE 19.—Index map of the Sheridan district.

shear zones in phonolite. The brecciated rocks of the shear zones have been altered and silicified; common alteration products are kaolin and sericite. Ore shoots at the Sheridan mine averaged 15 ounces per ton in silver. The ore minerals are not visible, and assays were necessary to determine ore-grade material.

At the American Flag mine, the silicified shear zones contained ore shoots composed of bornite, chalcopyrite, sphalerite, pyrite, and unidentified minute sulfides believed to be a silver mineral. At the Zalla M. and Silver Bell mines, silver is found in ore shoots of galena, chalcopyrite, and pyrite that occur in fissure veins composed of quartz, fluorite, and calcite. The veins range from 1 to 6 feet in thickness, are north trending, and steeply dipping. Although silver ore from the veins averaged around 15 ounces per ton in silver, and hand-sorted ore contained up to 118 ounces per ton, the main silver mineral of the district is unknown. Much of the silver appears to be contained in dark-gray streaks composed of minute opaque particles that might possibly be tetrahedrite or argentite. Some silver occurs in argentiferous galena, and Pearson (1967, p. 3) reports the presence of hessite (silver telluride) and krennerite (gold-silver telluride) in the Zalla M. vein.

Numerous prospect pits and short adits in the district expose quartz veins and pyritized shear zones; however, the veins and shear zones contain only traces of silver.

Production

Shipments of high-grade silver ore were made from the district as early as 1898; by 1919, the combined production from the Zalla M., Sheridan, Silver Bell, and American Flag mines reached $100,000. Except for small shipments in 1935, 1940, and 1968, mines of the district have been idle.

Principal Silver Deposits of Sheridan District

American Flag

Location: NE$\frac{1}{4}$ sec. 36, T. 38 N., R. 31 E.

Development: 97-foot drift with a 97-foot winze.

Geology: 2- to 8-foot-wide silicified shear zone in phonolite. Ore shoots up to 16 feet long average 16 ozs. per ton in silver, and 36 percent copper.

Ore minerals: Chalcopyrite, bornite, sphalerite, argentite and pyrite.

Production: 60 tons around 1918.

References: Patty, 1921, p. 196-198; Gage 1941, p. 172-173.

Sheridan

Location: SW$\frac{1}{4}$ sec. 24, T. 38 N., R. 31 E.

Development: Several adits with a total of around 900 feet of underground workings. Also a 300-foot shaft.

Geology: 1- to 7-foot-wide silicified shear zone in phonolite. Mill feed averaged 15 ozs. per ton in silver.

Ore minerals: Unidentified minute sulfides.

Production: 1910-1919: total production estimated between $60,000 and $100,000.

Reference: Patty, 1921, p. 194-196.

Silver Bell

Location: NW$\frac{1}{4}$SE$\frac{1}{4}$ sec. 25, T. 38 N., R. 31 E.

Development: 340-foot adit and a 600-foot drift.

Geology: Silicified shear zone in phonolite.
Two ore shoots contained up to 30 ozs.
per ton in silver.

Ore minerals: Unidentified minute sulfides.

Production: Early 1900's and 1940: minor
production

References: Huntting, 1956, p. 148.

Zalla M.

Location: NW¼ sec. 30, T. 38 N., R. 32 E.

Development: Two adits with 1,000 feet of
underground workings. Main crosscut
adit 470 feet long with 300-foot drift
on vein.

Geology: 1- to 6-foot-thick quartz vein in
phonolite. Secondary enrichment
from surface to depth of 150 feet.
Hand-sorted ore contained 86 to 118
ozs. per ton in silver, and 0.17 to
0.25 ozs. gold.

Ore minerals: Chalcopyrite, silver sulfides,
and pyrite.

Production: $40,000 prior to 1903. 1968:
300 tons.

Reference: Patty, 1921, p. 198-200.

NESPELEM DISTRICT

Location

The Nespelem district is in southeastern
Okanogan County, and extends from the Ferry County
border westward to the Okanogan River, and from the
north line of T. 32 N. to the Columbia River. Most
silver deposits of the district are in T. 31 N.,
R. 30 E., where elevations range from 1,815 feet at
Nespelem to 3,440 feet on the summit of Squaw
Mountain. The area is sparsely vegetated with sage-
brush and bunchgrass, and the northern slopes of most
mountains are covered by pine forests. Many parts of
the district, especially adjacent to the Columbia
River, are steep and rocky and have very little soil
cover.

Geology and Mineralization

Silver deposits of the Nespelem district occur
along shear zones, and in quartz fissure veins in gran-
odioritic rocks of the Colville batholith (Cretaceous).
The shear zones and quartz veins have general northwest strikes, and dip 40° to 70° northeast. They vary
from several inches to as much as 10 feet in thickness;
however, ore shoots in the shear zones average not
much more than 1 foot in thickness. The main silver
minerals of the shear zone deposits are sooty argentite,
stephanite, pyrargyrite, proustite, and native silver.
In unoxidized parts of the shear zones, stephanite,
galena, and sphalerite are the principal ore minerals.
In addition to the ore minerals, the shear zones contain sericitized granodiorite, quartz, calcite, fluorite,
rhodochrosite, and pyrite.

The ore shoots occur as tabular and lenticular
bodies that are separated from other parts of the shear
zones by black gouge. The size of the ore shoots is
generally small. The largest ore shoot at the Apache
mine was 12 to 18 inches thick, had a stope length of
40 feet, and a pitch length of 70 feet. At the Little
Chief mine, the most productive ore shoot had an average thickness of 2 feet, a stope length of 60 feet,
and a pitch length of 100 feet. In both of these mines
the richest ore occurred where minor cross fractures
intersected the main shear zones.

Crude ore from the district contained 14 to 25
ounces of silver per ton. Hand-sorted shipments of
ore from the Apache mine contained 560 ounces of
silver and 0.8 ounce of gold per ton. Ore from the
Little Chief contained 124 ounces per ton in silver,

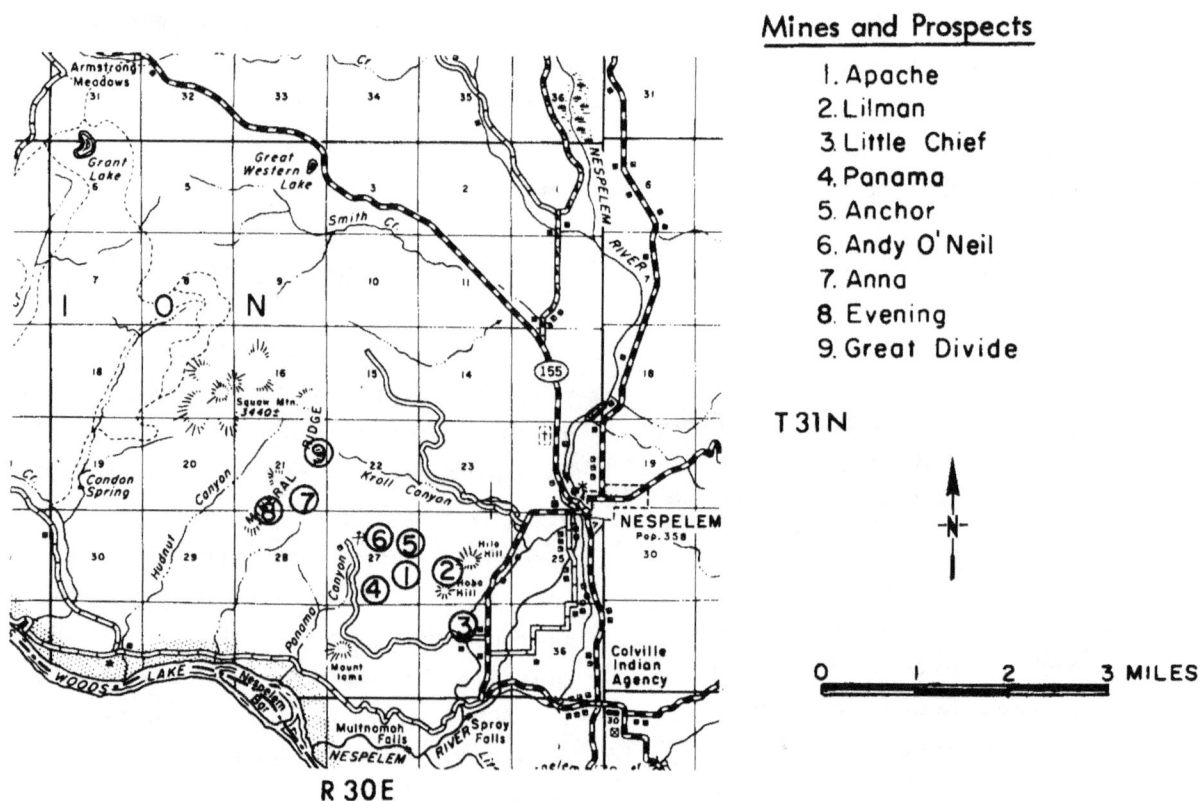

Mines and Prospects

1. Apache
2. Lilman
3. Little Chief
4. Panama
5. Anchor
6. Andy O'Neil
7. Anna
8. Evening
9. Great Divide

T 31 N

FIGURE 20.—Index map of the Nespelem district.

and 0.2 ounce in gold; however, some shipments assayed as much as 590 ounces per ton in silver. The richest ore shoots at producing mines in the district occurred beneath the leached zone, and for depths of 150 to 200 feet beneath the surface. From 15 to 40 feet beneath the surface most shear zones are leached and barren of ore minerals.

The quartz veins of the district have yielded only minor amounts of silver. The veins, which range from 1 to 4 feet in thickness, contain galena, sphalerite, tetrahedrite, chalcopyrite, and pyrite. For the most part the ore minerals are only sparsely disseminated in the quartz veins. In some veins, pods and lenses of ore minerals comprise up to 25 percent of the veins. Shipments of galena-rich ore from several quartz veins contained 1.2 to 10.3 ounces of silver

per ton, which indicates the low silver content of galena from the district.

In summary, some very rich silver ore has been mined in the Nespelem district; however, ore shoots are small and sporadically distributed in shear zones, which are the main metallized structures.

Production

Silver mining began in the Nespelem district in 1911, and by 1920 over 150,000 ounces of silver had been mined from high grade, near-surface deposits. The bulk of the production came from the Apache and Little Chief mines, which are 2 miles west of Nespelem. The total silver production for the district from 1911 through 1968, is estimated at 172,659 ounces. In addition to silver, mines of the district produced around 9,000 pounds of lead, 400 ounces of gold, and 300 pounds of copper. Currently (1975) there are no producing mines in the district.

Major Mines

Apache

This property, which is 1½ miles southwest of Nespelem, was the richest and most productive silver mine in the Nespelen district. The estimated production from 1911 to 1940 is around $250,000 in silver, and $20,000 in gold. Shipments to the Tacoma smelter in 1911, contained around $350 per ton in gold and silver. Small hand-sorted, high-grade shipments contained up to 2,444 ounces of silver to the ton. The bulk of the ore averaged 10 to 20 ounces of silver, and was concentrated at the Great Metals mill near Nespelem Falls, prior to shipment to the Tacoma smelter.

The mine is developed by a main haulage adit that is 700 feet long, and by 2,500 feet of underground workings on three levels. Most of the vein between the third level and the main adit has been stoped, and at least one stope extends from the main level to the surface. In addition to the main adit, the mine contains two shafts and three short adits, all of which follow a metallized shear zone.

The ore bodies occur along a shear zone, which strikes N. 50° E., and dips 50° NE., in granodiorite. In places the shear zone is displaced a foot or more by cross fractures. Some of the richest ore appears to have been deposited at the intersection of cross fractures and the main shear zone. The upper 25 feet of the deposit is oxidized, and leached of its metals. Below the oxidized zone, stephanite is the most important primary mineral, while argentite and native silver are the most important secondary minerals. Other ore minerals include galena, tetrahedrite, sphalerite, pyrargyrite, chalcopyrite, and pyrite. Common gangue minerals are quartz and rhodochrosite. The ore minerals occur in pods and lenses that pinch and swell along the strike and dip of the lode. The principal high-grade ore shoot, which has been mined out, had a maximum thickness of 2 feet, a stope length of 100 feet, and a pitch length of 150 feet. Other ore shoots had stope lengths of 30 to 40 feet. Much of this ore contained around 300 ounces of silver per ton, and almost 0.75 ounce of gold. In levels below the main adit, the gold and silver content of the ore decreased, whereas the copper and zinc content increased. Other parts of the lode are reported to have contained 16 to 20 ounces of silver per ton, over widths of 4 to 5 feet. Although the main ore shoots appear to have been mined out, further exploration of the shear zone below the zone of oxidation might possibly uncover additional shoots.

Little Chief

This property is three-quarters of a mile southeast of the Apache mine, and was the second most productive silver mine in the Nespelem district. Mining began in 1911, and by 1921 around $60,000 in silver had been produced from ore that contained 77 to 590 ounces of silver per ton, and 0.09 to 0.31 ounce of gold. The mine reportedly produced small amounts of ore in 1937 and 1954, which was concentrated at a 50-ton flotation mill near the portal of the main adit.

The main adit of the mine is around 560 feet long, and contains 640 feet of crosscuts and drifts.

Near the face of the adit a raise follows the lode for 200 feet to the surface. From near the surface to a depth of about 130 feet the lode has been mined for 60 feet on both sides of the raise. Ore was also mined from several small stopes in the main adit; however, the bulk of the production came from stopes in the raise.

Ore shoots occur in a well-defined shear zone that strikes N. 60° W. and dips 38° NE. The shear zone, in granodiorite, ranges from 2 to 10 feet in width, and consists of fragments of granodiorite, clay, sericite, quartz, and calcite. Ore minerals consist of argentite, pyrargyrite, stephanite, native silver, chalcopyrite, and pyrite. These minerals occur in small stringers, pods, and lenses along the shear zone. Within 100 feet of the surface, one lens of pyrargyrite, which was mined over a thickness of 2 feet and over a stope length of 30 feet, yielded around 1,000 ounces of silver. Although parts of the Little Chief workings contain sparsely disseminated ore minerals, no ore shoots appear to be present in the accessible mine workings. The owners of the mine report that ore is present in parts of the mine that are now inaccessible because of caved ground.

Principal Silver Deposits of Nespelem District

Anchor

Location: NE¼ sec. 27, T. 31 N., R. 30 E.
Development: Several caved adits and shafts.
Geology: Metallized shear zones in granodiorite. Select samples from the shear zones assayed up to 149.5 ozs. silver and 0.22 oz. gold.
Ore minerals: Argentiferous galena, sphalerite, and minor chalcopyrite.

Production: None.
Reference: Huntting, 1956, p. 300.

Andy O'Neil

Location: N¼ cor. sec. 27, T. 31 N., R. 30 E.
Development: 2 adits connected by a raise. Total underground workings about 800 feet.
Geology: 1- to 4-foot-thick quartz-calcite veins in granodiorite. Ore assays up to 30 ozs. silver and 20 percent lead.
Ore minerals: Galena, chalcopyrite, tetrahedrite, bornite, and pyrite.
Production: Minor production in 1965 and 1968.
Reference: Huntting, 1956, p. 300.

Anna

Location: SE¼ sec. 21, T. 31 N., R. 30 E.
Development: 140-foot adit and a 60-foot shaft.
Geology: 4-foot-thick shear zone in granodiorite contains metallized quartz veins, which assay up to 43 ozs. per ton in silver.
Ore minerals: Galena and pyrite with minor sphalerite and chalcopyrite.
Production: None.
Reference: Huntting, 1956, p. 300.

Apache

Location: Center E½ sec. 27, T. 31 N., R. 30 E.
Development: 2,500 feet of underground workings on three levels, including

three adits and three shafts. Main adit 560 feet long.

Geology: Tabular lenses and veinlets of ore along an 8- to 10-foot-wide shear zone in granodiorite. Ore shoots contained up to 300 ozs. per ton in silver and 3/4 oz. of gold; select high-grade ore contained up to 2,444 ozs. per ton in silver, and bulk of ore averaged 10 to 20 ozs. per ton in silver.

Ore minerals: Stephanite, argentite, pyrargyrite, galena, tetrahedrite, sphalerite, chalcopyrite, and pyrite.

Production: 1911-1940: $250,000 in silver and $20,000 in gold.

References: Pardee, 1918, p. 72-74, 68-70; Patty 1921, p. 207-209; Huntting, 1956, p. 300.

Evening

Location: S$\frac{1}{2}$SE$\frac{1}{4}$ sec. 21, T. 31 N., R. 30 E.

Development: Several small pits.

Geology: Lime-silicate bands in argillite and schist. Bands assay up to 20.2 ozs. per ton in silver.

Ore minerals: Galena, sphalerite, chalcopyrite, and pyrite.

Production: None.

Reference: Huntting, 1956, p. 302.

Great Divide

Location: E$\frac{1}{2}$NE$\frac{1}{4}$ sec. 21, T. 31 N., R. 30 E.

Development: 300-foot adit with 100 feet of crosscuts. Also several short adits.

Geology: Shear zone up to 50 feet wide in granodiorite contains metallized quartz veinlets. Ore minerals also occur in

pods up to 2 feet in diameter, which assay up to 100 ozs. per ton in silver.

Ore minerals: Galena, chalcopyrite, chalcocite, and pyrite.

Production: Unknown.

Reference: Huntting, 1956, p. 303.

Lilman

Location: SE$\frac{1}{4}$SE$\frac{1}{4}$ sec. 27, T. 31 N., R. 30 E.

Development: 35- and 50-foot shafts; 150-foot adit.

Geology: Lenses and veinlets of ore along a 1$\frac{1}{2}$- to 20-foot-wide shear zone in granodiorite. 3 inches of ore along hanging wall contained 80 ozs. per ton in silver and 0.20 oz. of gold.

Ore minerals: Tetrahedrite, chalcopyrite, pyrite, and pyrrhotite.

Production: 16 tons prior to 1940.

References: Huntting, 1956, p. 306.

Little Chief

Location: NE$\frac{1}{4}$NW$\frac{1}{4}$ sec. 35, T. 31 N., R. 30 E.

Development: 560-foot adit with 640 feet of drifts and crosscuts; 200-foot shaft with stopes.

Geology: 2- to 10-foot-wide shear zone in granodiorite containing stringers, pods, and lenses of ore minerals. Near-surface ore shoots were up to 2 feet thick and 30 feet long, and contained 77 to 590 ozs. of silver per ton, and 0.09 to 0.31 ozs. of gold.

Ore minerals: Argentite, pyrargyrite, steph-

anite, native silver, chalcopyrite,
and pyrite.

Production: 1911 to 1921: $60,000 in silver.
1937 and 1954: small amount of concentrates.

References: Patty, 1921, p. 213-214;
Pardee, 1918, p. 74; Huntting, 1956,
p. 306.

Panama

Location: Center S½ sec. 27, T. 31 N.,
R. 30 E.

Development: 30-foot crosscut adit with short
drift and small stope; 170-foot crosscut adit with 100-foot drift.

Geology: Quartz lenses and veins along 4-
foot-wide shear zone in granodiorite.
High-grade ore shoots contained 57
ozs. per ton in silver, and several
ozs. per ton in gold.

Ore minerals: Argentite, stephanite, native
silver, minor chalcopyrite, and pyrite.

Production: 1919, 1921, and 1936: 26 tons
total.

References: Patty, 1921, p. 209-211;
Huntting, 1956, p. 308.

PARK CITY DISTRICT

The Park City district is in east-central
Okanogan County and about 15 miles southwest of
Republic. Lead-silver deposits were discovered in
the district in 1898, and from 1905 to 1920 mining
was undertaken at several properties; however, total
production from the district has not exceeded $10,000.

The lead-silver veins of the district occur
along and near the contact between quartz monzonite
of the Colville batholith (Cretaceous) and Triassic
black shale, phyllite, and limestone. The veins are
of the quartz fissure type and have diverse strikes and
dips. They vary in thickness from less than 1 inch to
as much as 7 feet, and pinch and swell along their
strikes and dips. Most veins cannot be followed for
much more than 50 feet, but several veins are over a
hundred feet in length. Common sulfide minerals of
the veins include pyrite, galena, and sphalerite;
chalcopyrite and tetrahedrite are present only rarely.
Quartz is the chief gangue mineral and is accompanied
by minor calcite, fluorite, and sericite. Most commonly the sulfides are sparsely disseminated in the
quartz veins, but in some veins the sulfides are concentrated into small lenses, pods, and stringers. According to Pardee (1918, p. 95-102) the veins contain up to 182 ounces per ton in silver, traces to 0.10
ounce in gold, 2.7 to 40.3 percent lead, 0.05 to
0.3 percent copper, and 3.2 to 6 percent zinc.
Although small shipments of ore, containing 30 percent lead and 20 ounces of silver per ton, were made
years ago, exploration has failed to discover any
large silver ore shoots in the district. Most lead-
silver ore shoots appear to be small and isolated.

Principal Silver Deposits of Park City District

Eureka

Location: NW¼NW¼ sec. 7 and SW¼SW¼
sec. 6, T. 33 N., R. 32 E.

Development: 200-foot adit.

Geology: 1½-foot-thick quartz vein in black
shale. Select samples assayed up to
150.6 ozs. silver and 41.4 percent
lead.

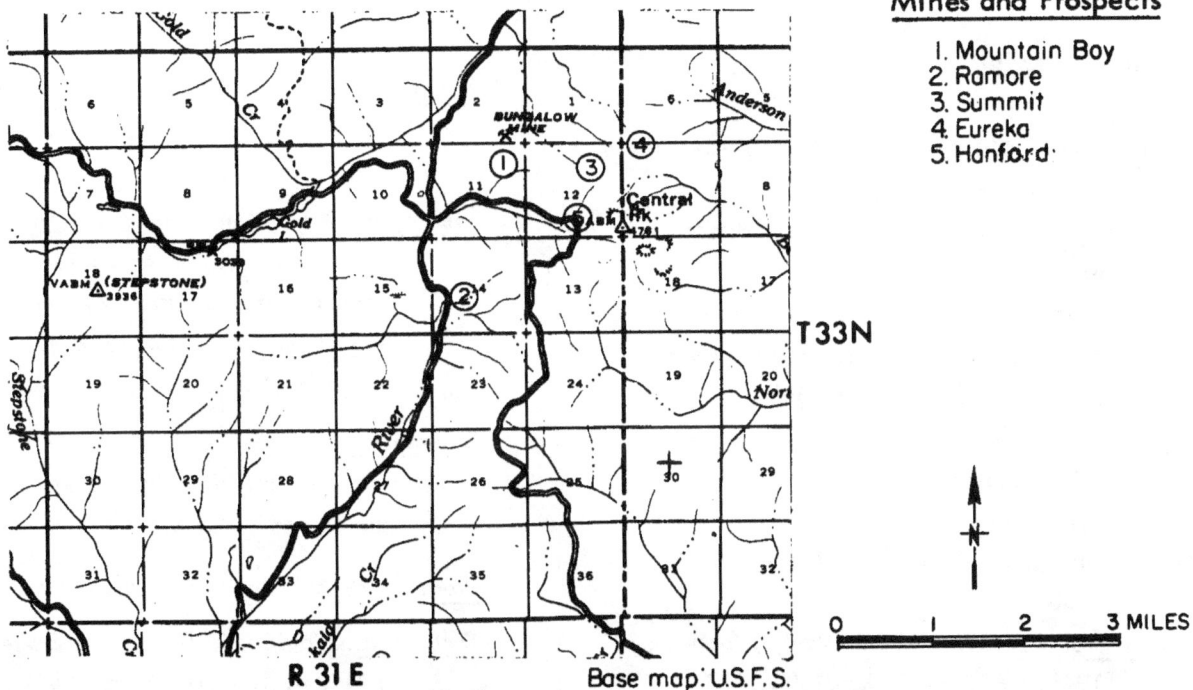

FIGURE 21.—Index map of the Park City district.

Ore minerals: Galena, chalcopyrite, and
    pyrite.
Production: None.
Reference: Huntting, 1956, p. 218.

Hanford

Location: S¼ cor. sec. 12, T. 33 N.,
    R. 31 E.
Development: Inclined shaft.
Geology: 2- to 3-foot-thick quartz vein in
    argillite. Select samples contained
    25.9 percent lead and 20.48 ozs.
    silver per ton.
Ore minerals: Galena and pyrite.
Production: None.
Reference: Huntting, 1956, p. 219.

Mountain Boy

Location: NE¼ sec. 11, T. 33 N., R. 31 E.
Development: Four adits, one of which is
    155 feet long.
Geology: Quartz lenses up to 1½ feet thick
    along shear zones in argillite and
    limestone. Select samples of ore con-
    tained 7 to 182 ozs. per ton in silver.
Ore minerals: Galena, sphalerite, chalcopy-
    rite, tetrahedrite, chalcocite, covel-
    lite, and pyrite.
Production: Four to five carloads prior to
    1918 averaged $60 per ton in lead and
    silver.
References: Pardee, 1918, p. 94-95; Gage,
    1941, p. 183-185; Patty, 1921,
    p. 192.

Ramore

Location: NE$\frac{1}{4}$SW$\frac{1}{4}$ sec. 14, T. 33 N.,
R. 31 E.

Development: 440-foot adit with a 600-foot
crosscut and a 45-foot winze.

Geology: Five-foot quartz vein with disseminated sulfides along shear zone in
granodiorite. Assays of ore show up
to 59 ozs. per ton in silver, 9.6 percent lead, and 0.07 oz. gold.

Ore minerals: Galena, sphalerite, tetrahedrite, chalcopyrite, pyrite, and
pyrrhotite.

Production: Small shipments in the early
1900's.

References: Pardee, 1918, p. 98-99, Gage,
1941, p. 188-190; Bancroft, 1914,
p. 207-209.

Summit

Location: NE$\frac{1}{4}$ sec. 12, T. 33 N., R. 31 E.

Development: 1,000-, 350-, 150-, and
100-foot adits.

Geology: Quartz lenses along a 1$\frac{1}{2}$-foot-
wide shear zone in granodiorite. Ore
shoots contained up to 90 ounces of
silver per ton, 0.12 ounce gold, and
40 percent lead.

Ore minerals: Galena, sphalerite, chalcopyrite, tetrahedrite, arsenopyrite,
and pyrite.

Production: 50 tons of galena around 1910.

References: Pardee, 1918, p. 94; Gage,
1941, p. 183-185; Patty, 1921,
p. 190-191.

OTHER SILVER DISTRICTS OF EASTERN WASHINGTON

Thus far the discussion of the silver districts of eastern Washington has been confined to the Okanogan Highlands of northeastern Washington. Elsewhere in eastern Washington, silver deposits occur in the Cascade Mountains. Although silver is known to occur at several hundred mines and prospects east of the crest of the Cascades, only a few deposits contain silver in notable amounts. These deposits are in the Meadow Creek and Stehekin districts of Chelan County and the Cle Elum and Gold Creek districts of Kittitas County. In other mining districts silver appears to be minor and accompanies gold, lead, zinc, and copper.

CHELAN COUNTY

Chelan County ranks third in the production of silver by counties; however, the bulk of the silver has come from copper mining operations. From 1904 through 1969, the county produced 2,822,225 ounces of silver valued at $2,364,048. The largest silver producer was the Holden mine that produced 2,147,744 ounces of silver from 1938 through 1957. In addition to silver, the mine produced around 216 million pounds of copper from a total of 10.6 million tons of ore. The second largest silver producer was the Gold King mine, which from 1949 to 1967 produced 420,000

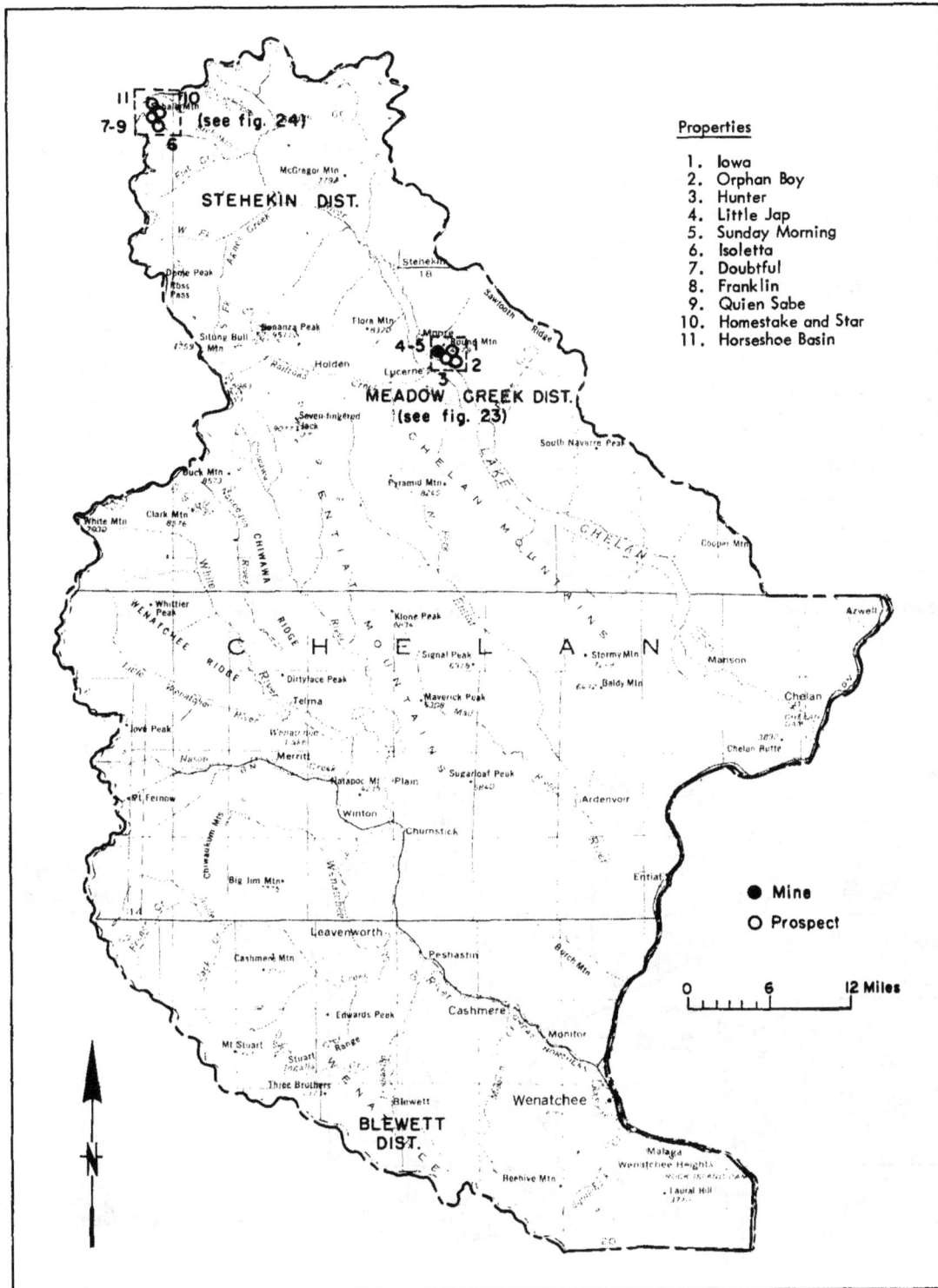

Properties

1. Iowa
2. Orphan Boy
3. Hunter
4. Little Jap
5. Sunday Morning
6. Isoletta
7. Doubtful
8. Franklin
9. Quien Sabe
10. Homestake and Star
11. Horseshoe Basin

● Mine
○ Prospect

0   6   12 Miles

Base from U.S.G.S. State of Washington
1:500,000, 1962

FIGURE 22.—Silver deposits of Chelan County.

ounces of silver from around 1 million tons of ore. In addition to silver the mine produced 367,500 ounces of gold. An additional 14,860 ounces of silver came from copper mining operations at the Red Mountain mine, which operated from 1935 through 1937. Prior to 1935, the production of silver in Chelan County was insignificant. Almost yearly since 1890, several hundred ounces of silver have come from small gold mining operations, mainly through the refining of gold bullion.

Aside from the silver that occurs in the copper and gold ores, only two districts of Chelan County appear to have anomalous concentrations of silver. Quartz veins in the Meadow Creek district are sparsely metallized with silver minerals, whereas in the Stehekin district silver occurs in argentiferous galena. The production of silver, as well as other metals from these districts, has been insignificant.

MEADOW CREEK DISTRICT

This district is near the head of Lake Chelan and lies northeast of the lake. The terrain is steep and rocky and elevations range from 1,096 feet on the shores of the lake to 8,500 feet on the highest peaks above the lake. The district is not yet accessible by roads.

The ore minerals occur in lenticular quartz veins that have as their host rocks pre-Upper Cretaceous biotite gneiss. The veins, which are up to 4 feet thick, occur in hydrothermally altered shear zones as much as 7 feet thick. The ore minerals of the veins are galena, sphalerite, chalcopyrite, tetrahedrite, pyrargyrite, and native silver. These ore minerals are almost always accompanied by pyrite, pyrrhotite, and arsenopyrite. Deposits that are richest in silver occur near the shore of Lake Chelan; at least

Mines and Prospects

1. Hunter
2. Iowa
3. Little Jap
4. Orphan Boy
5. Sunday Morning

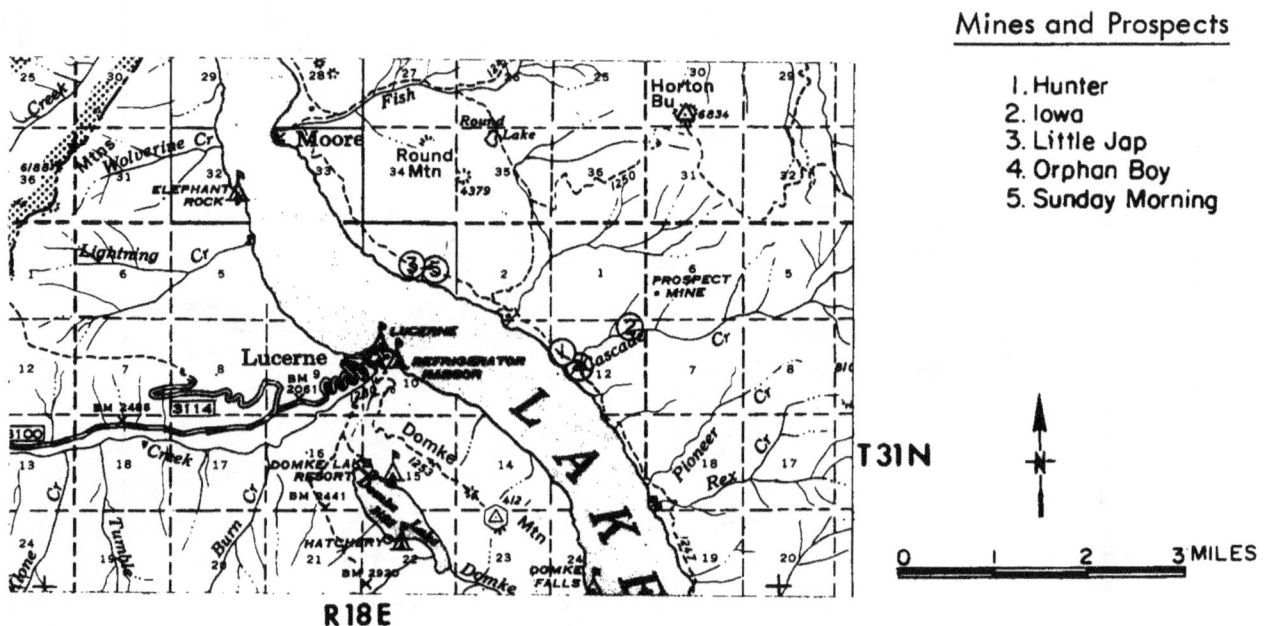

FIGURE 23.—Index map of the Meadow Creek district.

one deposit is now under water, due to the raising of the lake level by the dam at Chelan. Deposits near the lake consist of narrow quartz veins that contain rich seams of pyrargyrite, tetrahedrite, and galena. As much as 140 ounces of silver per ton has been reported for some seams. Although early reports mention the presence of silver minerals, Webb (1957, p. 33-47) did not observe any silver minerals in the veins of the district. Sulfides noted by Webb were arsenopyrite, pyrrhotite, marcasite, chalcopyrite, and sphalerite.

In addition to silver-bearing quartz veins, several of the district's copper-metallized quartz-feldspar dikes carry silver. Up to 21 ounces of silver per ton have been reported in seams as irregular masses of solid sulfides that occur along the walls of the dikes. Common sulfide minerals include chalcopyrite, pyrrhotite, and pyrite, which when oxidized, impart a distinct yellowish-brown color to the surface parts of the dikes. Most dikes crop out on the mountainside several thousand feet above the lake level.

Principal Silver Deposits of Meadow Creek District

Hunter

Location: NW¼ sec. 12, T. 31 N., R. 18 E.
Development: 15-foot adit.
Geology: Quartz vein with 4-inch paystreak containing up to 140 ozs. silver and 1 oz. gold per ton.
Ore minerals: Pyrargyrite and tetrahedrite.
Production: None.
Reference: Huntting, 1956, p. 206.

Iowa

Location: NE¼ sec. 12, T. 31 N., R. 18 E.

Development: 40-foot adit.
Geology: Quartz vein with 1-foot paystreak that contains up to 200 ozs. silver and 3 ozs. gold per ton.
Ore minerals: Unknown.
Production: None.
Reference: Huntting, 1956, p. 113.

Little Jap

Location: NW¼SE¼ sec. 3, T. 31 N., R. 18 E.
Development: Main adit below lake level; 30-foot adit 200 feet above lake level.
Geology: Quartz veinlets in granite gneiss.
Ore minerals: Pyrargyrite, chalcopyrite, pyrite, and pyrrhotite.
Production: None.
Reference: Huntting, 1956, p. 286.

Orphan Boy

Location: SW¼ sec. 12, T. 31 N., R. 18 E.
Development: 50-foot adit and a 10-foot inclined shaft.
Geology: Quartz veins up to 3 inches thick in a 7-foot-wide altered shear zone. Sample of footwall assayed 28.64 ozs. silver, 0.02 oz. gold, and 0.11 oz. zinc. Sample of hanging wall assayed 4.64 ozs. silver, 0.13 percent lead, and 0.16 percent zinc.
Ore minerals: Pyrite, pyrrhotite, galena, sphalerite, arsenopyrite, and minor pyrargyrite.
Production: None.
Reference: Huntting, 1956, p. 286.

Sunday Morning

Location: NW¼SE¼ sec. 3, T. 31 N.,
R. 18 E.

Development: 70-foot adit near lake level.

Geology: 4-foot-wide altered shear zone in
granite gneiss contains sparsely
metallized quartz veinlet.

Ore minerals: Galena, pyrargyrite, and
pyrite.

Production: 2-ton shipment prior to 1897
grossed $250 per ton.

Reference: Huntting, 1956, p. 287.

STEHEKIN DISTRICT

The Stehekin district is in the northwestern
corner of Chelan County, and much of the district is in
the North Cascades National Park, which is closed
to mineral entry. In the 1890's, work was undertaken
on several small high-grade argentiferous galena deposits, which resulted in the production of 2,200

pounds of ore that had a net value of $60 per ton.
To date (1975) most work in the district has been undertaken on the Horseshoe Basin property; however,
the property has been idle since 1952.

The richest argentiferous galena deposits are
in Horseshoe Basin near the crest of the Cascades.
Elevations in the basin range from 3,200 to 7,200
feet. Quartz fissure veins predominate and have as
their host rocks pre-Jurassic biotite, quartz diorite,
and hornblende gneisses, many of which are migmatitic. The veins occupy well-developed shear
zones in the gneiss, and have general east to northeast strikes and near-vertical dips. The veins range
from several inches to 7 feet in thickness; most are
only sparsely metallized with argentiferous galena,
sphalerite, pyrite, and minor chalcopyrite. Parts of
some veins contain lenses of solid galena up to 3 feet
in thickness, and assay as much as several hundred
ounces per ton in silver. In general, the ore minerals
occur only as small isolated pods and lenses. Sphalerite appears to be more abundant than galena and
chalcopyrite, while pyrite, pyrrhotite, and arsenopyrite, in order of decreasing abundance, make up the
bulk of the sulfide in the veins. The Davenport vein,

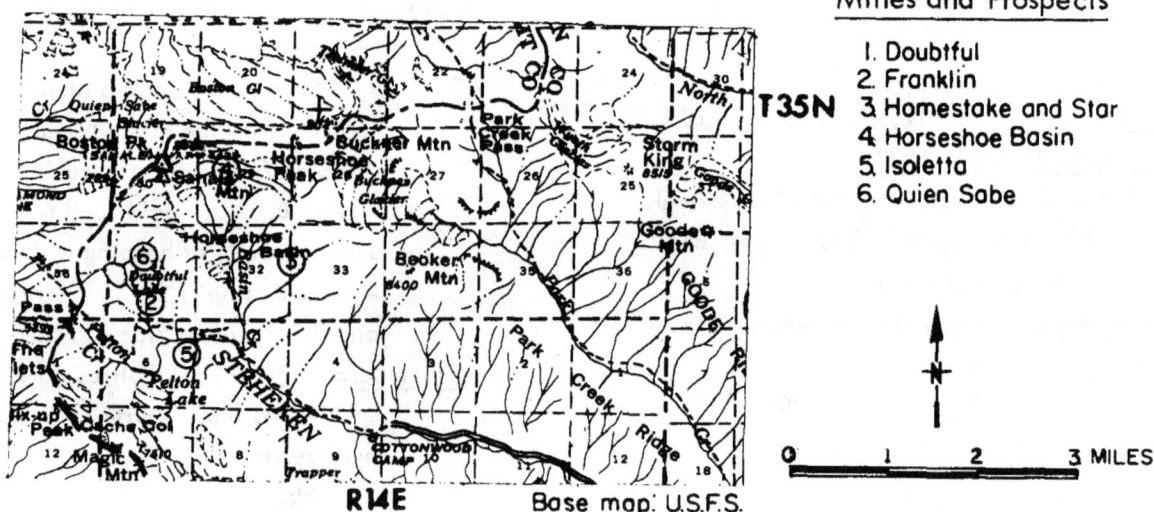

Mines and Prospects

1. Doubtful
2. Franklin
3. Homestake and Star
4. Horseshoe Basin
5. Isoletta
6. Quien Sabe

FIGURE 24.—Index map of the Stehekin district (Horseshoe Basin area).

which is the most persistent vein in the basin, contains an ore shoot 2½ to 9 feet thick, and over 200 feet in stope lengths. Parts of the ore shoot contain 40 percent lead and 52 ounces per ton in silver. Huntting (1956, p. 207) reports the average metal content of the ore as follows: 4.7 percent lead, 4.8 percent zinc, 0.35 percent copper, 1.78 percent arsenic, 0.062 ounce per ton in gold, and 9.8 ounces per ton in silver.

Principal Silver Deposits
of Stehekin District
(Horseshoe Basin Area)

Doubtful

Location: SE¼NW¼ sec. 31, T. 35 N.,
R. 14 E.
Development: 30- and 100-foot adits.
Geology: 16- to 24-inch quartz vein in
gneiss. Vein contains a 4-inch band
of sulfides, which assays 87.5 to
100 ozs. silver per ton and 54 to 62
percent lead.
Production: None.
Reference: Huntting, 1956, p. 206.

Franklin

Location: NE¼SW¼ sec. 31, T. 35 N.,
R. 14 E.
Development: Unknown.
Geology: 4-foot-thick quartz vein, select
samples of which assayed 89 to 100
ozs. silver per ton and 42 to 54 per-
cent lead.
Ore minerals: Galena and pyrite.
Production: None.
Reference: Huntting, 1956, p. 206.

Homestake and Star

Location: SW¼NW¼ sec. 33, T. 35 N.,
R. 14 E.
Development: 30-foot open cut.
Geology: 4-foot-wide ore body, select
samples of which assayed 112 to 400
ozs. silver per ton and 1 oz. gold.
Ore minerals: Unknown.
Production: None.
Reference: Huntting, 1956, p. 285.

Horseshoe Basin

Location: N½ sec. 29 and NE¼ sec. 32,
T. 35 N., R. 14 E.
Development: 1,000-foot crosscut adit with
1,000 feet of drifts.
Geology: Persistent shear zone in granodiorite can be followed for 1,060 feet
along strike. Ore shoot 3 feet wide
and 330 feet long averages 9.8 ozs.
silver, 0.062 oz. gold, 4.7 percent
lead, 4.8 percent zinc, 0.35 percent
copper, and 1.78 percent arsenic.
Ore minerals: Galena, sphalerite, chalcopyrite, arsenopyrite, and pyrite.
Production: None.
Reference: Huntting, 1956, p. 206.

Isoletta

Location: SW¼NW¼ sec. 5, T. 34 N.,
R. 14 E.
Development: 215-foot adit.
Geology: Unknown. Select samples assayed
300 to 700 ozs. silver per ton.
Ore minerals: Unknown. Probably argentif-
erous galena.
Production: 1 ton in the early 1900's.
Reference: Huntting, 1956, p. 286.

Quien Sabe

Location: SE¼NW¼ sec. 31, T. 35 N.,
R. 14 E.

Development: 250-foot drift.

Geology: Quartz vein, select samples of
which assayed 103 to 204 ozs. silver
and 0.1 oz. gold.

Ore minerals: Galena, chalcopyrite, arsenopyrite, and pyrite.

Production: None.

Reference: Huntting, 1956, p. 207.

KITTITAS COUNTY

Silver has been reported at many metal occurrences in Kittitas County, but the county has never
had a major silver-producing mine. From 1884 to

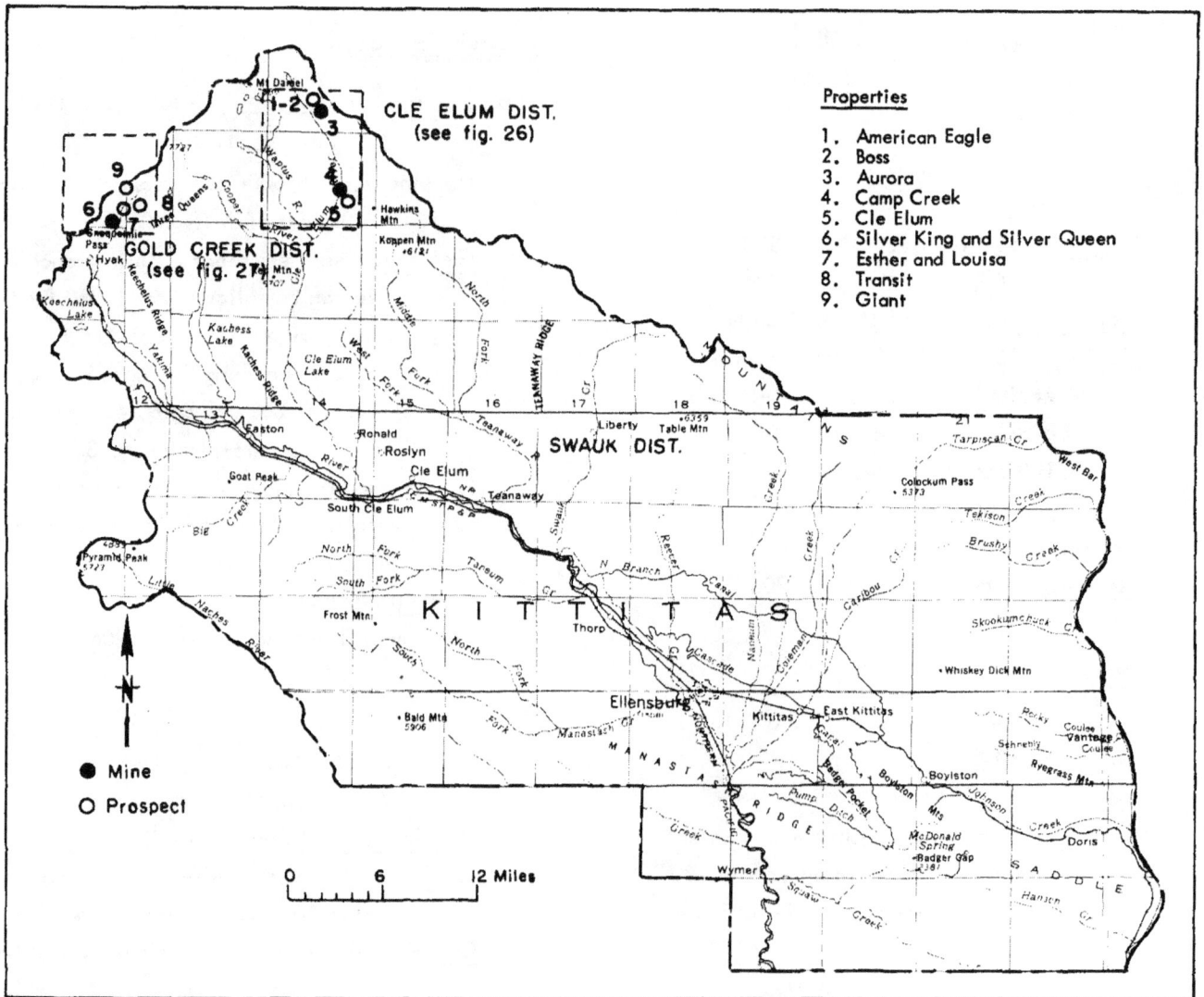

FIGURE 25.—Silver deposits of Kittitas County.

1970, mines of the county produced around 25,000 ounces of silver, almost all of which came from the refining of placer gold that was mined in the Swauk district. Lode gold mines in the Swauk district also produced silver, but the production was minor. In the Cle Elum and Gold Creek districts, silver was a byproduct of gold, copper, and lead mining operations; however, the combined metal production from mines of these districts was less than $10,000. Currently no lode mines are operating in the county, and only small-scale placer mining is taking place in the Swauk district.

Although many mines in the Cle Elum and Gold Creek districts report the presence of silver, only at a few properties does the silver content of the veins exceed 10 ounces per ton. Most silver deposits appear to be small and consist of sparsely metallized shear zones and quartz veins in andesite and granitic rocks. The andesite is part of the Keechelus Andesite (Oligocene and Eocene), whereas the granitic rocks are part of the Mount Stuart batholith (Cretaceous). Common ore minerals of metallized shear zones and quartz veins are pyrite, pyrrohotite, arsenopyrite, chalcopyrite, galena, and sphalerite. Free gold, pyrargyrite, and tetrahedrite are rarely present.

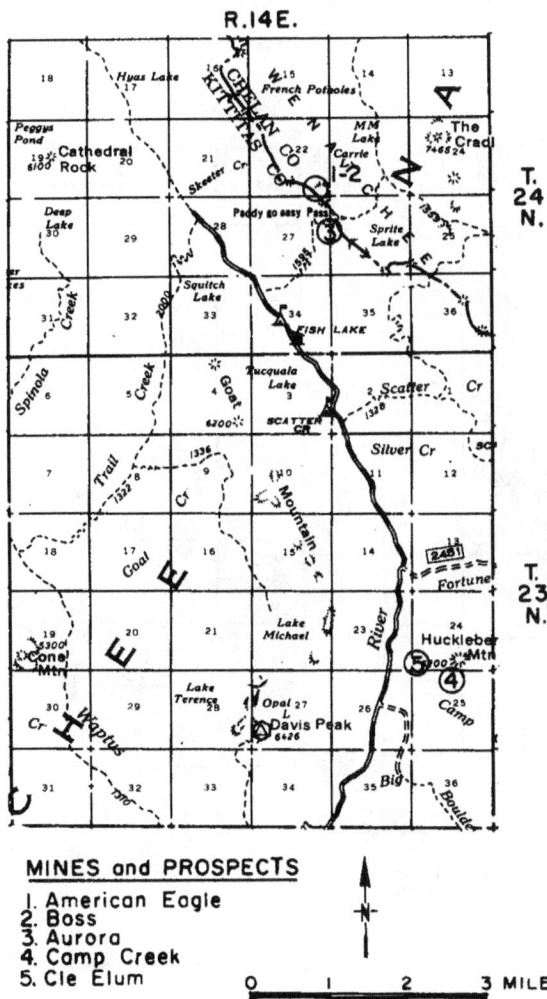

FIGURE 26.—Index map of the Cle Elum district.

MINES and PROSPECTS
1. American Eagle
2. Boss
3. Aurora
4. Camp Creek
5. Cle Elum

CLE ELUM DISTRICT

Principal Silver Deposits

American Eagle

Location: SE¼SE¼ sec. 22, T. 24 N., R. 14 E.

Development: Unknown.

Geology: 6-foot vein with 1½ feet of ore that assays up to 1.5 ozs. gold and 15 ozs. silver per ton.

Ore minerals: Unknown.

Production: None.

Reference: Bethune, 1892, p. 133.

Aurora

Location: Secs. 26 and 27, T. 24 N., R. 14 E.

Development: 2 shafts, each over 200 feet deep.

Geology: 4-foot-thick quartz vein. Select samples of vein assayed 1 oz. gold,

14 ozs. silver, 6 percent copper, and
28 percent arsenic.

Ore minerals: Free gold and arsenopyrite.

Production: Minor.

Reference: Huntting, 1956, p. 130.

Boss

Location: SE¼SE¼ sec. 22, T. 24 N.,
R. 14 E.

Development: Unknown.

Geology: 6-foot thick vein with 1½ feet of
ore that assays 1.5 ozs. gold and 20
ozs. silver per ton.

Ore minerals: Unknown.

Production: None.

Reference: Bethune, 1892, p. 134.

Camp Creek

Location: Near N¼ cor. sec. 25, T. 23 N.,
R. 14 E.

Development: 40-, 60-, and 160-foot adits.

Geology: Narrow quartz veins in greenstone.
Ore from veins carried 42 ozs. silver
and 0.14 oz. gold per ton.

Ore minerals: Galena, chalcopyrite, sphalerite, and pyrite.

Production: 15 tons.

Reference: Huntting, 1956, p. 298.

Cle Elum

Location: Near SW. cor. sec. 24, T. 23 N.,
R. 14 E.

Development: 700-foot inclined shaft.

Geology: 4-foot vein, select samples of
which assay up to 55 ozs. silver and
0.2 oz. gold.

Ore minerals: Pyrite.

Production: None.

Reference: Huntting, 1965, p. 298.

GOLD CREEK DISTRICT

Principal Silver Deposits of Gold Creek District

Esther and Louisa

Location: N½NE¼ sec. 27, T. 23 N.,
R. 12 E.

Development: 2 adits.

Geology: Quartz vein in andesite. Vein

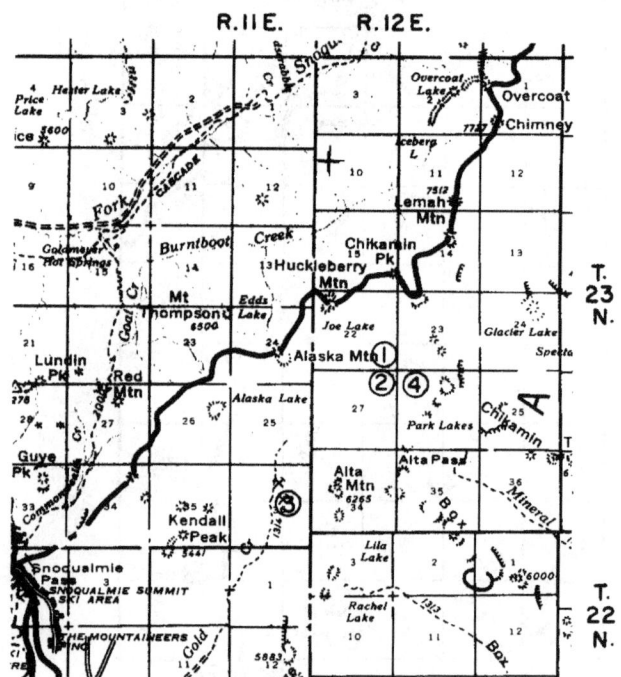

MINES and PROSPECTS
1. Giant
2. Esther and Louisa
3. Silver King and Silver Queen
4. Transit

FIGURE 27.—Index map of the Gold Creek district.

sparsely metallized but high-grade
ore assayed 240 ozs. silver and 0.5
oz. gold per ton.

Ore minerals: Pyrargyrite, galena, sphalerite, and pyrite.

Production: Minor production in 1896.

Reference: Huntting, 1956, p. 298.

Giant

Location: SE¼ sec. 22, T. 23 N., R. 12 E.

Development: 500-foot adit with short drifts
and a 100-foot adit with a winze.

Geology: ½- to 6-foot-wide metallized shear
zones in granitic rocks. Ore minerals
make up less than 5 percent of the
shear zones, and assays show up to
14.96 ozs. silver, 0.14 oz. gold,
0.08 percent copper, and 0.16 percent
lead. Select dump samples assayed
12.40 ozs. silver, 2.40 oz. gold,
0.3 percent copper, and 2.1 percent
lead.

Ore minerals: Galena, chalcopyrite, and
pyrite.

Production: Unknown.

Reference: Gualtieri and others, 1973,
p. 64-66.

Silver King and Silver Queen

Location: Near center sec. 36, T. 23 N.,
R. 11 E.

Development: Caved adit and inclined shaft.

Geology: Quartz vein in andesite assays up
to 22.4 ozs. silver, 0.07 oz. gold,
and 0.063 percent copper.

Ore minerals: Unknown.

Production: None.

Reference: Gualtieri and others, 1973,
p. 68-69.

Transit

Location: NW¼NW¼ sec. 26, T. 23 N.,
R. 12 E.

Development: 100-foot adit and a 60-foot
drift with a winze.

Geology: Narrow metallized shear zones in
andesite. Sample from the shear zone
assayed 10.70 ozs. silver, 0.08 oz.
gold, and 0.56 percent copper.

Ore minerals: Pyrite.

Production: None.

Reference: Gualtieri and others, 1973,
p. 66-67.

PART III

SILVER OCCURRENCES
of
WESTERN WASHINGTON

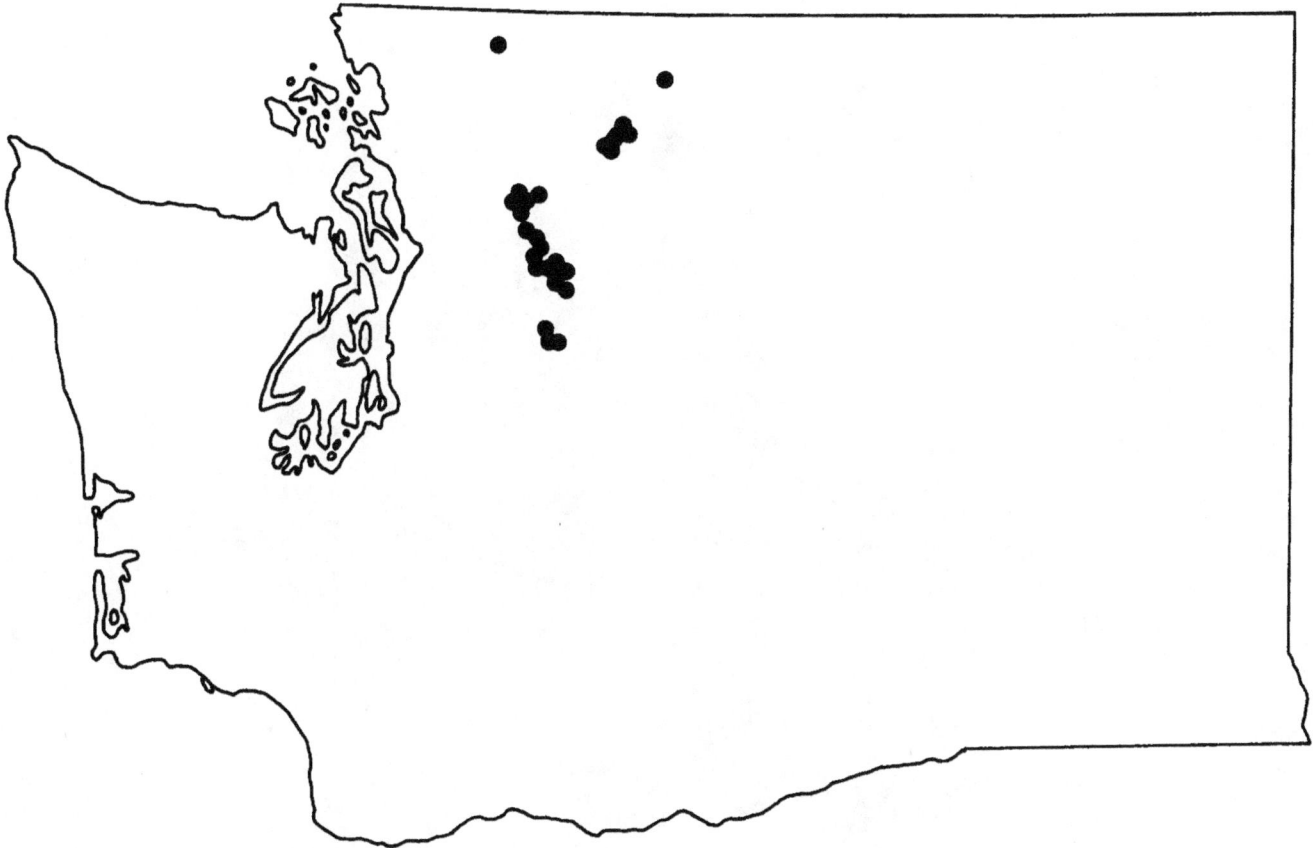

CASCADE MOUNTAINS

INTRODUCTION

No major silver districts are present in western Washington; from 1890 through 1969, western counties produced only 2,255,585 ounces of silver valued at around $1,888,745. From 1903 through 1956, Snohomish County produced 31,099 ounces. The combined pro- Whatcom County produced 57,837 ounces, and King County produced 34,503 ounces. The combined production of Skagit, Pierce, Lewis, and Skamania Counties is only around 3,000 ounces. Silver from Snohomish County was produced as a byproduct of the gold and copper mining operations in the Monte Cristo and Index districts; silver from Whatcom County was mainly a byproduct of gold mining operations in the Mount Baker district; in King County silver was a byproduct of gold mining operations in the Miller River district.

Currently (1975), no metal mines are operating in western Washington. However, in recent years exploration has been undertaken on several large low-grade copper deposits. If placed into production, appreciable amounts of silver would be produced as a byproduct of copper mining operations.

PHYSIOGRAPHY

The principal silver deposits of western Washington are in the parts of Whatcom, Skagit, Snohomish, and King Counties that fall within the north half of the Cascade Mountains physiographic province. This is an area of extremely rugged terrain resulting from alpine glaciation. Throughout much of the area, serrated rocky ridges and rocky pinnacles of the alpine type of topography dominate the landscape. Much of the terrain has elevations in excess of 6,000 feet, which is the general timberline. Above timberline, many small lakes occupy glacial cirques, and along rocky ridges glacial tarns are common. Above 7,000 feet elevation, snowfields and glaciers may be found throughout the year. Mount Baker (10,778 feet) and Glacier Peak (10,436 feet), which are Pleistocene volcanoes, are the highest mountains in the Northern Cascades. Numerous streams occur throughout the region and many of them occupy deep, glaciated valleys. Major rivers include the Snoqualmie, Snohomish, Skykomish, Stillaguamish, Skagit, and Nooksack. Up to 120 inches of rain per year on the western slopes of the Northern Cascades produce thick stands of Douglas fir, western hemlock, and western red cedar. Undergrowth is dense, and consists of alder, willow, maple, salmonberry, blackberry, ferns, and devil's club. Above timberline the forests give way to typical alpine vegetation, consisting of groves of mountain hemlock and alpine fir surrounded by meadows of heather, huckleberry, and blueberry.

GENERAL GEOLOGY AND MINERALIZATION

The Northern Cascade Mountains are composed of a core of granitic rocks and metamorphosed sedimentary and volcanic rocks, which are flanked by younger sedimentary and volcanic rocks (fig. 6).

The older metamorphosed sedimentary and volcanic rocks consist mainly of pre-Jurassic gneiss, schist, quartzite, phyllite, marble, and greenstone. The post-Jurassic sedimentary rocks consist chiefly of shale, arkose, graywacke, and conglomerate. Post-Jurassic volcanic rocks are mainly andesite, basalt, and rhyolite flows with associated pyroclastic rocks. The volcanic cones of Mount Baker and Glacier Peak are composed of andesite flows and pyroclastic rocks of Pleistocene age; however, ash eruptions from Mount Baker occurred as recently as 1843, and currently (1975) a vent near the top of the mountain is emitting steam and gases. The last major eruption from Glacier Peak was 12,000 years ago.

The granitic rocks, which make up part of the core of the Cascade Mountains, were intruded into the older rocks of the Cascades during Cretaceous and early Tertiary times. Most Cretaceous granitic rocks are from 100 to 120 million years old, whereas the granitic Tertiary intrusive rocks are as young as 14 to 20 million years. The largest granitic masses, which are of batholithic size, are composed mainly of quartz diorite and granodiorite. The smaller granitic masses consist of stocks and plugs of Tertiary age, which vary in composition from granite to quartz diorite. The borders of most granitic intrusive rocks are gradational into the enclosing rocks; however, the borders of some granitic masses contain intrusive breccias.

The rocks of the Cascade Mountains have been tightly folded along north-northwesterly-trending axes. This folding developed early, as is shown by foliation and folding in the older metamorphic rocks, which make up part of the core of the Cascades. The Tertiary rocks of the region have also been folded and faulted, though not to the extent of the older rocks. The northern part of the Cascade Mountains has been uplifted more than the southern part so that the overall structure of the Cascades resembles a south-plunging anticline. The uparching of the Cascades occured

over a time period of about 6 million years that began in Pliocene time and continued into Pleistocene. Because of the greater uplifting, and subsequent erosion in the north Cascades, many mineral deposits have been exposed. In the south Cascades, where uplifting and erosion has not been as great, great thicknesses of Tertiary volcanic rocks predominate.

Most silver deposits of the Northern Cascades of western Washington are fissure veins or mineralized joints and shear zones that occur in or close to bodies of granitic rocks (see fig. 5). Common host rocks for the deposits include schist, argillite, quartzite, phyllite, granodiorite, diorite, granite, andesite, and porphyry. The veins range from a fraction of an inch to as much as 15 feet in thickness, with the average thickness being around 3 feet. Pinching and swelling of veins is widespread. Common gangue minerals of silver-bearing veins include quartz and calcite that are accompanied by gouge and wallrock fragments. Common ore minerals include pyrite, pyrrhotite, arsenopyrite, chalcopyrite, galena, and sphalerite. Among the less common minerals are tetrahedrite, chalcocite, bornite, molybdenite, stibnite, argentite, pyrargyrite, native silver, and free gold. The bulk of the silver is contained in argentiferous galena and argentian tetrahedrite. Only rarely are silver sulfides and native silver visible in the silver ores. However, ruby silver was reported to be a major mineral at the "45" mine in Snohomish County. Although select samples from some silver deposits contained as much as 600 ounces per ton in silver, the average silver content of the silver veins is only around 6 ounces per ton. Several mines have shipped small amounts of ore that contained 20 to 40 ounces per ton in silver. At the "45" mine the average silver content of ore shipped was 93 ounces per ton, while 300,000 tons of ore mined at Monte Cristo averaged 6 ounces per ton in silver. In western Washington the ground-water level almost coincides with the surface. In most deposits oxidation of the primary sulfides rarely extends deeper than 10 feet, and secondary enrichment of the silver deposits does not exist.

An exception to the vein-type silver deposit is the silver-bearing breccia at the Great Excelsior mine in Whatcom County. At this property a rather large body of pyritized and silicified breccia in andesite, argillite, and slate contains 3.4 ounces per ton in silver and 0.09 ounce in gold.

SNOHOMISH COUNTY

Snohomish County leads in the production of silver for western Washington counties; however, almost all the metal was a byproduct of gold and copper mining operations. From 1890 until 1907, gold mines in the county produced the bulk of the silver. After 1907, most silver came from copper-mining operations.

TABLE 23.—Peak production years for silver in Snohomish County[1]

Years	Mines	Total production (ounces)
1890-1907	Monte Cristo Sunset "45"	1,796,200
1916-1920	Sunset	34,000
1923-1930	Sunset	110,000
1939-1943	Iowa Florence Rae Sunset Kromona	60,000

[1] Production prior to 1903 cannot be substantiated. The figures shown are believed to be realistic and are based upon the most reliable data available.

The production of silver in Snohomish County came to an end in 1969, when the Kromona copper mine in the Sultan district ceased operations. Peak years for the production of silver in Snohomish County are shown in table 23.

Although silver-bearing veins occur in all mining districts of Snohomish County, no district is classed as a silver district, because the value of gold or copper produced in the past has exceeded that of silver. Only one mine in the county can be classed as a silver mine, this being the "45" mine in the Sultan Basin district. From 1896 to 1901, the mine produced 3,185 tons of ore that averaged 93 ounces per ton in silver and 0.74 ounce in gold. The gross value of the ore was around $100,000.

The richest silver-bearing veins in the county are found in the Sultan Basin, Silver Creek, Silverton, and Darrington districts; however, the deposits appear to be small. The silver occurs mainly in argentiferous galena that for the most part is sparsely disseminated in quartz veins and shear zones. In deposits that have been mined, galena, as well as other ore minerals, occurred in lenses and stringers that were erratically distributed in quartz veins and shear zones. Common sulfides, other than galena, include pyrite, pyrrhotite, chalcopyrite, sphalerite, and arsenopyrite. The less common ore minerals of the veins are bornite, tetrahedrite, scheelite, pyrargyrite, native silver, and molybdenite. Common host rocks for the silver deposits are argillite, slate, and phyllite of possible Triassic-Jurassic age, as well as granodiorite and quartz diorite of mid-Tertiary age. Although select samples from some deposits contain up to 200 ounces per ton in silver, the average silver content of most deposits is less than 5 ounces per ton. The average silver content of ore from the Monte Cristo district was 6 ounces per ton; the gold content of the ore was 0.6 ounce per ton. At the Sunset mine, where the ore averaged around 2.45 percent copper, the average

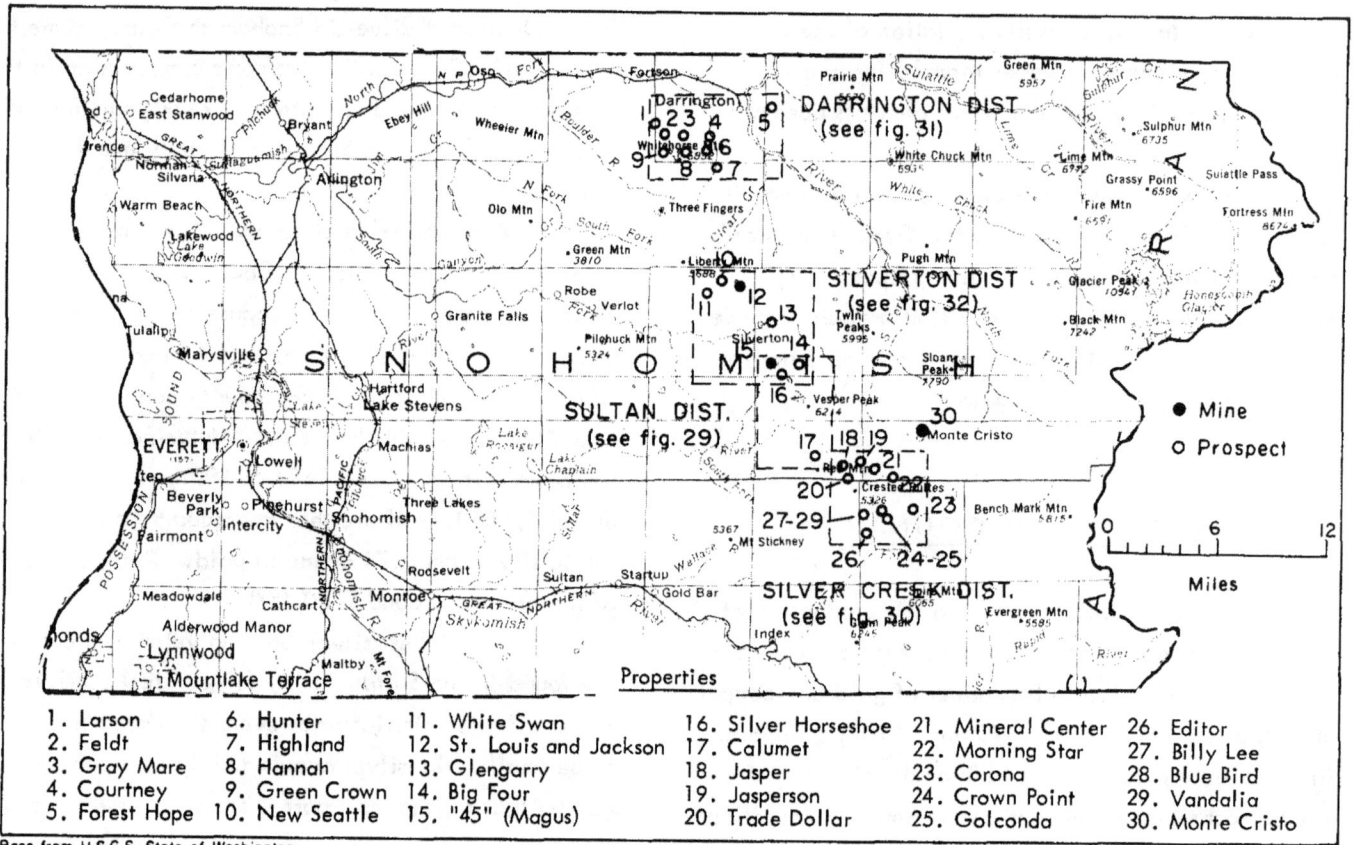

FIGURE 28.—Silver deposits of Snohomish County.

silver content was only 0.59 ounce per ton, and the gold content was 0.0057 ounce.

Major Mines

"45" (Magus) Mine

This property, which is at the headwaters of Williamson Creek in the Sultan district, is the only major silver mine in western Washington. As early as 1896, ore containing up to 135 ounces per ton in silver was shipped to the Everett smelter, and by 1901, the mine produced 3,185 tons of ore valued at $99,255. Ore was shipped that contained 48.4 to 171.4 ounces per ton in silver, and averaged 93 ounces. Based on partial smelter returns the "45" mine produced around 300,000 ounces of silver, and 2,356 ounces of gold. In the early 1900's, a 12,000-foot-long, single-rope aerial tramway extended from the mine to Silverton.

The main mine workings comprise a lower No. 3 adit (2,400 feet long), an intermediate No. 2 adit (650 feet long), and an upper No. 1 adit (300 feet long). Stopes extend from the upper adit to 120 feet beneath the No. 2 adit, for a vertical distance of 320 feet. As far as is known, the entire production from the mine came from these stopes.

The "45" vein is a quartz calcite fissure vein that occurs in a northwest-trending, steeply dipping shear zone in argillite, quartzite, and schist. The vein is from 6 inches to 6 feet in width, and appears to be at least 3,000 feet long. It carries pyrite, arsenopyrite, pyrrhotite, scheelite, chalcopyrite, galena, and tetrahedrite. According to early reports ruby silver (pyrargyrite or proustite) was present in the upper workings of the mine, where high-grade ore shoots varying from a few inches to 5 feet in thickness were mined. Assays of veins on the Magus and Hard to Beat claims show 6 to 18 percent arsenic, 4 percent zinc, 4.6 to 6.5 percent lead, 8 to 10.4 ounces of silver, and 0.28 to 0.6 ounce of gold. The bulk of the high-grade silver ore in the mine appears to have been removed; however, ore containing 8 to 10 ounces per ton in silver is still present. The possibility also exists that underdeveloped veins that parallel the "45" vein may contain high-grade ore shoots.

SULTAN DISTRICT

Principal Silver Deposits

Calumet

Location: Near center of $N\frac{1}{2}$ sec. 27, T. 29 N., R. 10 E.

Development: 250-foot adit.

Geology: $\frac{1}{2}$- to 2-foot thick quartz vein in metamorphic rocks. Vein assays trace to 1.03 oz. gold, 0.56 to 5.33 ozs. silver, 0.48 to 6.5 percent copper, and 0.10 to 4.85 percent zinc.

Ore minerals: Galena, chalcopyrite, sphalerite, pyrite, and pyrrhotite.

Production: None.

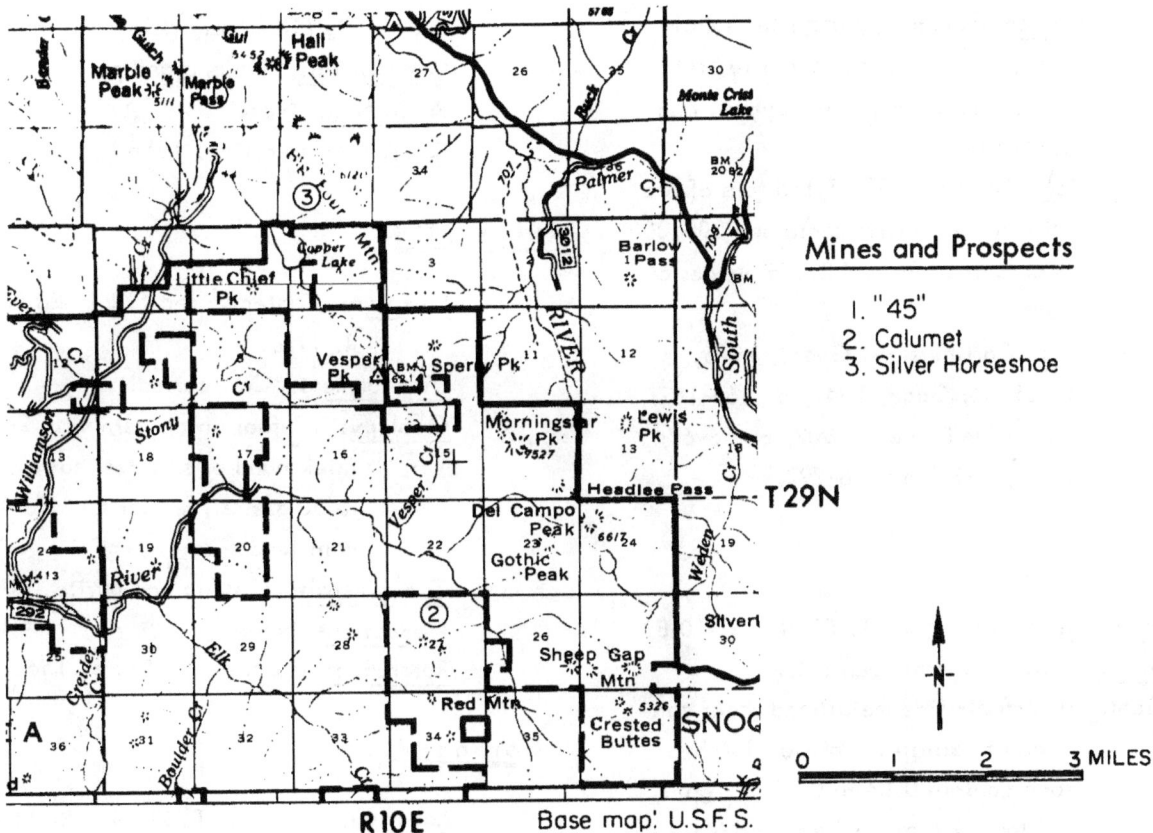

Mines and Prospects

1. "45"
2. Calumet
3. Silver Horseshoe

FIGURE 29.—Index map of the Sultan district.

Reference: Huntting, 1956, p. 157.

"45" (Magus)

Location: S½ sec. 29, T. 30 N., R. 10 E.

Development: Over 4,000 feet of underground workings, and several large stopes. 300-, 700-, and 2,400-foot adits have developed the deposit over a vertical distance of 350 feet.

Geology: Metallized shear zones in schist, argillite, and quartzite. Main shear zone averages 24 inches in width, and appears to be at least 3,000 feet long. High grade ore contained up to 171 ounces per ton in silver. Average ore contained 93 ounces per ton in silver, and 0.74 ounce of gold.

Ore minerals: Galena, pyrargyrite, sphalerite, chalcopyrite, scheelite, tetrahedrite, arsenopyrite, pyrite, and pyrrhotite.

Production: 1896 to 1902, 3,185 tons of ore shipped to smelters contained 300,000 ounces of silver and 2,356 ounces of gold.

References: Carithers and Guard, 1945, p. 39-46; Gage, 1941, p. 166-167; Landes and others, 1902, p. 66-67; Patty, 1921, p. 296-297.

Silver Horseshoe

Location: SW¼ sec. 33, T. 30 N., R. 10 E.

Development: Several open cuts.

Geology: 2-foot-wide metallized shear zone in quartz diorite. Parts of the shear zone contain 0.04 to 0.42 oz. gold, 4.8 to 160.8 ozs. silver, and 3.6 to 11.3 percent lead.

Ore minerals: Sparse galena, sphalerite, chalcopyrite, scheelite, molybdenite, pyrite, pyrrhotite, and arsenopyrite.

Production: None.

Reference: Huntting, 1956, p. 322.

SILVER CREEK DISTRICT

Principal Silver Deposits

Billy Lee

Location: NW¼SE¼ sec. 7, T. 28 N., R. 11 E.

Development: 25- and 112-foot adits.

Geology: 18-foot-thick vein with 2-foot-thick band of sulfides. Said to assay high in silver, gold, and lead.

Ore minerals: Galena and pyrite.

Production: None.

Reference: Huntting, 1956, p. 233.

Blue Bird

Location: Center S½ sec. 7, T. 28 N., R. 11 E.

Development: Crosscut adit.

Geology: 27-foot-thick vein with 2-foot-thick band of sulfides that assays up to 50 ozs. silver per ton and 48 percent lead.

Ore minerals: Galena and pyrite.

Production: None.

Reference: Huntting, 1956, p. 156.

Corona

Location: Sec. 10, T. 28 N., R. 11 E.

Development: 65-foot adit.

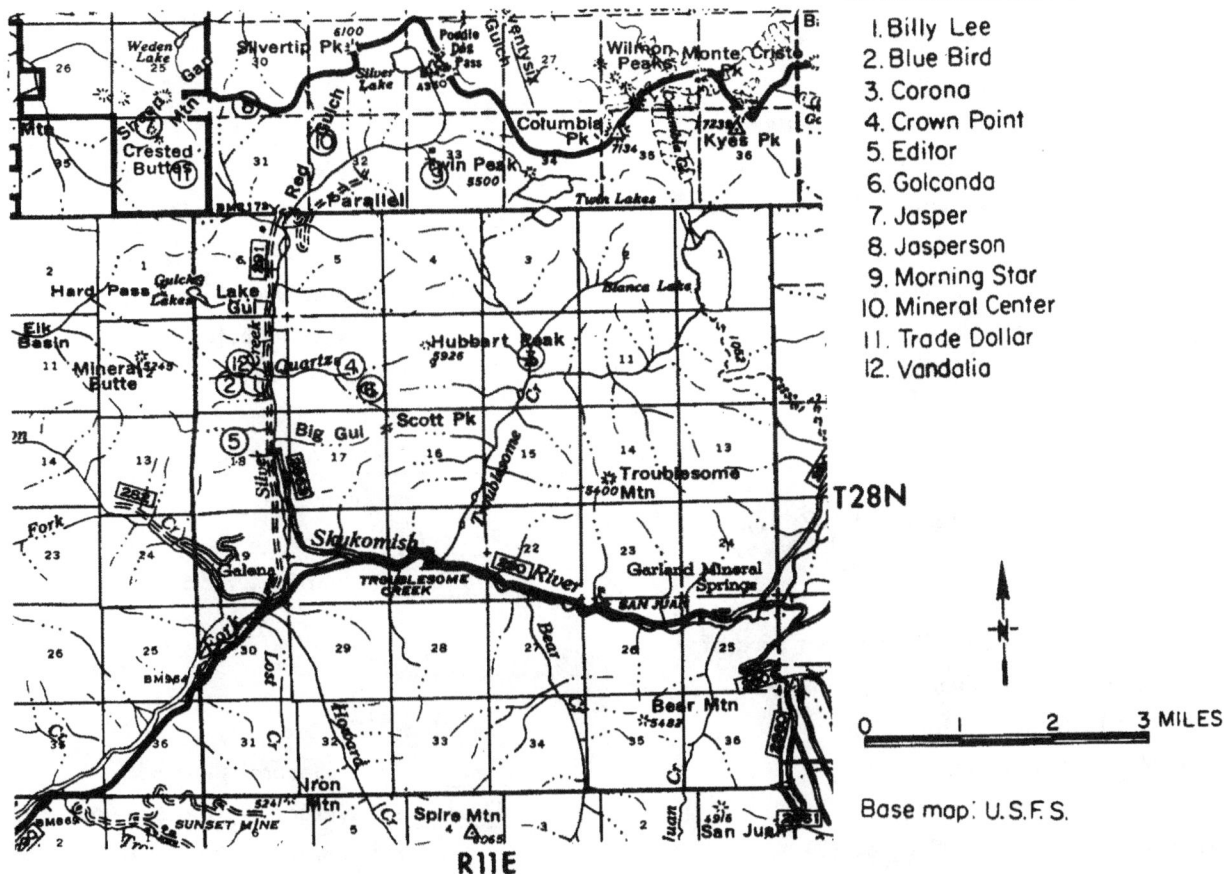

FIGURE 30.—Index map of the Silver Creek district.

Mines and Prospects

1. Billy Lee
2. Blue Bird
3. Corona
4. Crown Point
5. Editor
6. Golconda
7. Jasper
8. Jasperson
9. Morning Star
10. Mineral Center
11. Trade Dollar
12. Vandalia

Geology: 6-foot-thick vein with a 3-foot-
thick band of ore that assays up to
70 ozs. silver and 1 oz. gold.

Ore minerals: Unknown.

Production: None.

Reference: Huntting, 1956, p. 318.

Crown Point

Location: Sec. 8, T. 28 N., R. 11 E.

Development: 30-foot adit.

Geology: Several veins ranging from 4 to 20
feet in thickness. Assays of 107 ozs.
silver and 40 percent lead reported.

Ore minerals: Galena and pyrite.

Production: None.

Reference: Huntting, 1956, p. 233.

Editor

Location: Sec. 18, T. 28 N., R. 11 E.

Development: Unknown.

Geology: 2-foot-thick vein carrying up to
30 ozs. silver per ton.

Ore minerals: Galena.

Production: None.

Reference: Huntting, 1956, p. 318.

Golconda

Location: SE¼ sec. 8, T. 28 N., R. 11 E.

Development: 40-foot adit.

Geology: 4- to 20-foot-wide metallized shear zone with 2-foot band of ore that assays 42 percent lead and high in silver and gold.

Ore minerals: Galena and pyrite.

Production: None.

Reference: Huntting, 1956, p. 234.

Jasper

Location: NE¼NW¼ sec. 36, T. 29 N., R. 11 E.

Development: 27-foot adit.

Geology: 7-foot vein with a 1-foot band of ore. Select dump sample assayed 138 ozs. silver and 44 percent lead.

Ore minerals: Galena.

Production: None.

Reference: Huntting, 1956, p. 234.

Jasperson

Location: SE¼SW¼ sec. 31, T. 29 N., R. 11 E.

Development: 3,000 feet of workings in several adits.

Geology: Narrow metallized fracture zones in granodiorite. Ore shoots are 3 to 8 inches wide and sporadic. Eight samples assayed trace to 19.20 ozs. silver, 0.02 to 0.44 oz. gold, trace to 4.5 percent copper and 0.8 to 15.1 percent lead.

Ore minerals: Galena, sphalerite, stibnite, vanadinite, cinnabar, chalcopyrite, arsenopyrite, and pyrite.

Production: None.

Reference: Huntting, 1956, p. 160.

Mineral Center

Location: W½ sec. 26, T. 29 N., R. 11 E.

Development: Total of 3,500 feet of underground workings in 3 adits.

Geology: Metallized shear zones up to 40-feet wide in metasediments. Samples from one shear zone assayed 7.38 ozs. silver and 0.10 oz. gold.

Ore minerals: Abundant pyrite and minor chalcopyrite.

Production: None.

Reference: Huntting, 1956, p. 162.

Morning Star

Location: SW¼ sec. 33, T. 29 N., R. 11 E.

Development: Several short adits.

Geology: Several metallized shear zones with one ore shoot 6 feet wide that assays up to 41 ozs. silver and 40 percent lead.

Ore minerals: Galena, pyrite, and chalcopyrite.

Production: None.

Reference: Huntting, 1956, p. 235.

Trade Dollar

Location: SE¼ sec. 36, T. 29 N., R. 10 E.

Development: 50-foot drift.

Geology: 4-foot vein, select samples of which assay 63 percent lead and 37 ozs. silver.

Ore minerals: Galena and pyrite.

Production: None.

Reference: Huntting, 1956, p. 236.

Vandalia

Location: SW¼ sec. 7, T. 28 N., R. 11 E.

Development: 220-foot adit and a 75-foot
shaft with 2 levels.

Geology: 20-foot-wide shear zone with ½ to
3 feet of ore. Select samples assayed
up to 300 ozs. silver, 2 ozs. gold,
and 40 percent lead.

Ore minerals: Galena and pyrite.

Production: None.

Reference: Huntting, 1956, p. 236.

DARRINGTON DISTRICT

Principal Silver Deposits

Courtney

Location: SW¼ sec. 27, T. 32 N., R. 9 E.

Development: 10-foot adit.

Geology: 3-foot vein containing up to 15 ozs.
silver, 4 percent lead, and 0.5 oz.
gold.

Ore minerals: Unknown.

Mines and Prospects

1. Courtney
2. Feldt
3. Forest Hope
4. Gray Mare
5. Green Crown
6. Hannah
7. Highland
8. Hunter
9. Larson

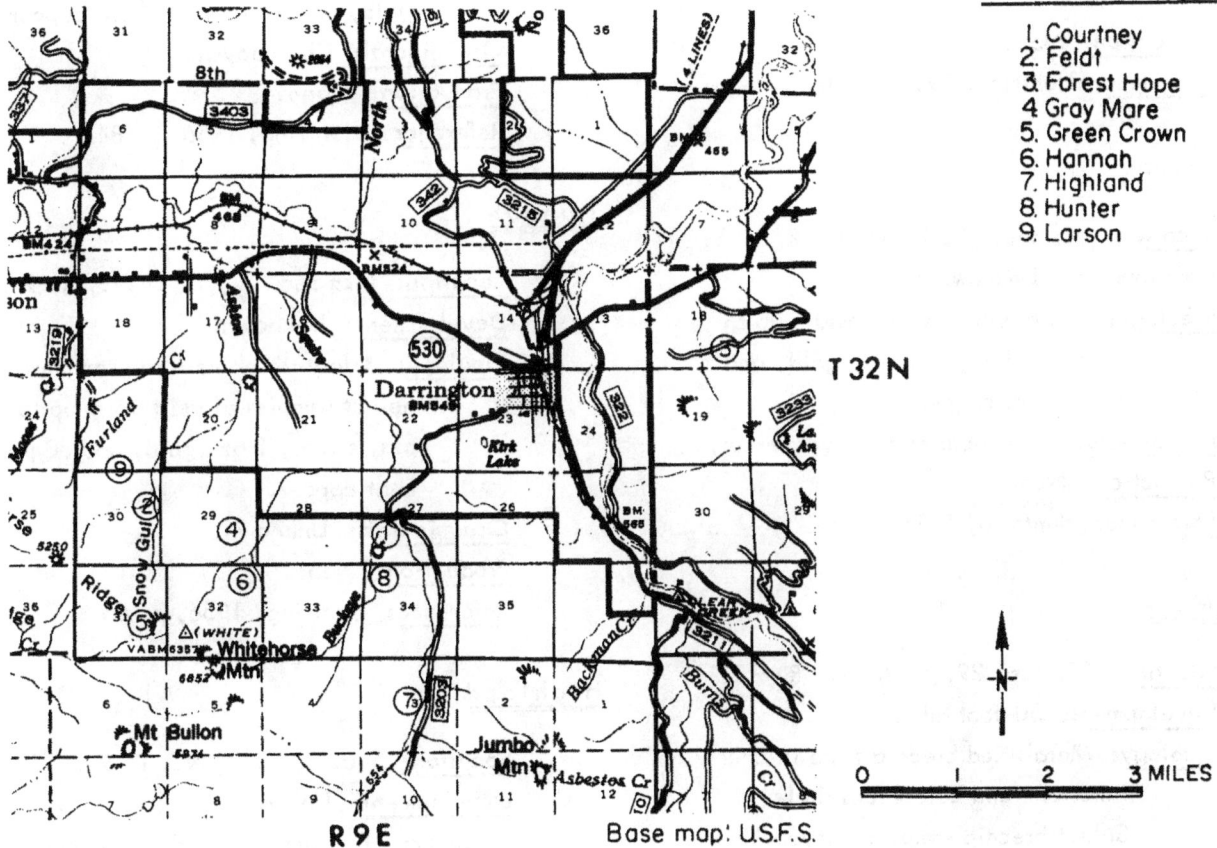

FIGURE 31.—Index map of the Darrington district.

Production: None.
Reference: Huntting, 1956, p. 82.

Feldt

Location: SE¼NE¼ sec. 30, T. 32 N.,
    R. 9 E.
Development: 2 open cuts.
Geology: Metallized breccia along fault zone
    in altered volcanic rocks. Ore min-
    erals occur in a 7-inch zone in the
    breccia.
Ore minerals: Stibnite, berthierite, chalco-
    pyrite, galena, sphalerite, tetrahed-
    rite, cobaltite, pyrite, and arsenopy-
    rite.
Production: None.
Reference: Huntting, 1956, p. 318.

Forest Hope

Location: SE¼ sec. 18, T. 32 N., R. 10 E.
Development: Unknown.
Geology: 8-foot vein, which assays up to
    18 ozs. silver, 1.5 ozs. gold, and
    18 percent copper.
Ore minerals: Cinnabar and native mercury.
Production: None.
Reference: Huntting, 1956, p. 83.

Gray Mare

Location: SE¼ sec. 29, T. 32 N., R. 9 E.
Development: 50-foot adit.
Geology: Metallized breccia zone in cherty
    limestone and calcareous shale.
    Select breccia samples assayed 15.8
    to 245 ozs. silver, 20 to 70 percent
    lead, 1 to 3 percent zinc, 0.3 to

9.8 percent antimony, 0.01 oz. gold,
    and 0.34 percent arsenic.
Ore minerals: Galena, sphalerite, and silver
    sulfides.
Production: None.
Reference: Huntting, 1956, p. 319.

Green Crown

Location: SE¼ sec. 31, T. 32 N., R. 9 E.
Development: Unknown.
Geology: 100-foot-wide shear zone contain-
    ing metallized quartz veins up to 10
    inches thick. Parts of some veins
    assay up to 32 ozs. silver, 5 ozs.
    gold, and 26 to 42 percent copper.
Ore mineral: Chalcopyrite.
Production: None.
Reference: Huntting, 1956, p. 84.

Hannah

Location: NE¼ sec. 32, T. 32 N., R. 9 E.
Development: Unknown.
Geology: 3-foot-thick vein in granodiorite.
    Select samples of vein assay up to 41
    ozs. silver, 1 oz. gold, and 30 per-
    cent copper.
Ore minerals: Unknown.
Production: None.
Reference: Huntting, 1956, p. 85.

Highland

Location: Sec. 3, T. 31 N., R. 9 E.
Development: Unknown.
Geology: 18 inches of ore that assays 15 to
    40 ozs. silver and 1.8 percent copper.
Ore minerals: Unknown.

Production: None

Reference: Huntting, 1956, p. 85.

Hunter

Location: NW$\frac{1}{4}$ sec. 34, T. 32 N., R. 9 E.

Development: 50-foot adit.

Geology: 3-foot vein with a 9-inch band of ore that assays 8 to 40 ozs. silver, 1 oz. gold, 10 percent copper, and 4 percent lead.

Ore minerals: Galena, chalcopyrite, and pyrite.

Production: None.

Reference: Huntting, 1956, p. 85.

Larson

Location: Secs. 19 and 30, T. 32 N., R. 9 E.

Development: Unknown.

Geology: Veins in granite and limestone assay up to 25 ozs. silver, 0.1 oz. gold, 60 percent lead, and 3 percent zinc.

Ore minerals: Galena and pyrite.

Production: None.

Reference: Huntting, 1956, p. 235.

SILVERTON DISTRICT

Principal Silver Deposits

Big Four

Location: NW$\frac{1}{4}$ sec. 33, T. 30 N., R. 10 E.

Development: 100-foot adit.

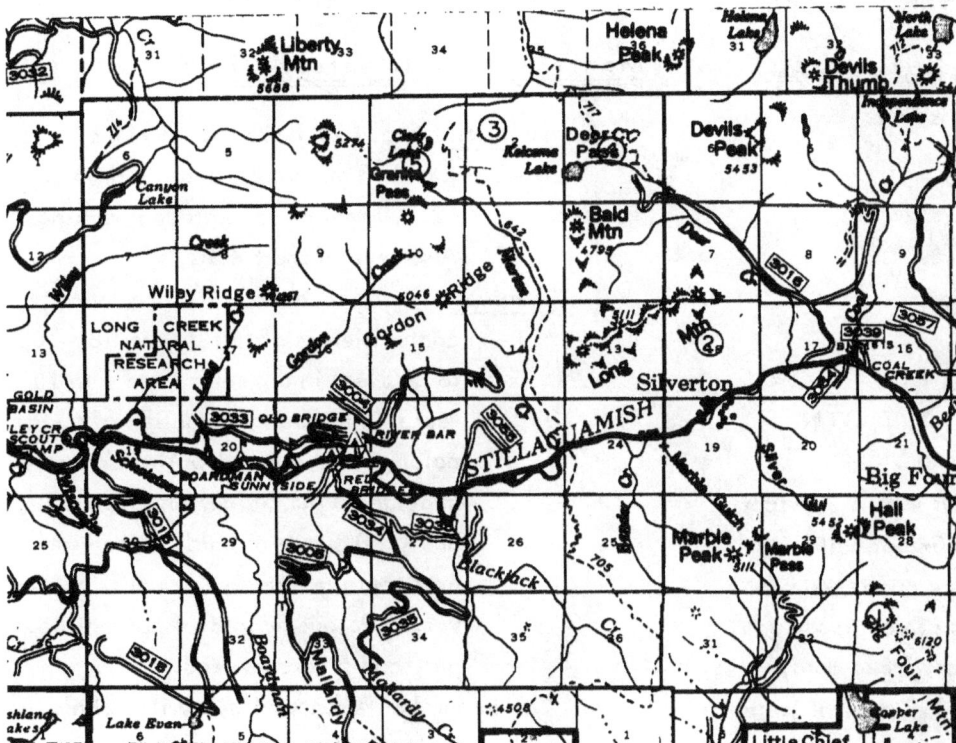

Mines and Prospects

1. Big Four
2. Glengarry
3. New Seattle
4. St. Louis and Jackson
5. White Swan

FIGURE 32.—Index map of the Silverton district.

Geology: 30-inch metallized zone at contact
between granite and syenite. Assays
of up to 20 ozs. silver per ton reported.
Ore minerals: Galena and arsenopyrite.
Production: None.
Reference: Huntting, 1956, p. 233.

Glengarry

Location: Sec. 18, T. 30 N., R. 10 E.
Development: Unknown.
Geology: 45 inches of "ore," with assays of
up to 150 ozs. per ton in silver.
Ore mineral: Tetrahedrite.
Production: None.
Reference: Huntting, 1956, p. 319.

New Seattle

Location: NW¼ sec. 2, T. 20 N. R. 9 E.
Development: Short adit.
Geology: 5-foot vein, parts of which assay
350 ozs. per ton in silver.
Ore mineral: Tetrahedrite.
Production: None.
Reference: Huntting, 1956, p. 321.

St. Louis and Jackson

Location: Near center sec. 1, T. 30 N.,
R. 9 E.
Development: 600-foot drift with a 200-foot
crosscut. Also a 150-foot adit.
Geology: 1- to 6-foot-thick quartz vein in
granodiorite. Parts of vein assay up
to 30 percent copper, 1 oz. gold,
25 ozs. silver, and 1.2 percent tungsten.
Ore minerals: Chalcopyrite, scheelite, tet-

rahedrite, and pyrite.
Production: Minor in the early 1900's.
Reference: Huntting, 1956, p. 90.

White Swan

Location: Sec. 3, T. 30 N., R. 9 E.
Development: 50-foot adit.
Geology: Unknown, but vein reported to
carry up to 200 ozs. per ton in silver.
Ore minerals: Unknown.
Production: None.
Reference: Huntting, 1956, p. 323.

MONTE CRISTO DISTRICT

Principal Silver Deposit

Monte Cristo (Mystery, Pride)

Location: Sec. 22, T. 29 N., R. 11 E.
Development: 3 main adits contain around
12,000 feet of underground workings.
Also several short adits.
Geology: Shear zones in schist and tonalite
contain lenses of sulfide ore from 100
to 300 feet in diameter, and 1 to 15
feet thick. Gangue consists of quartz,
calcite, and sheared wall rock. Ore
averaged 5 ounces per ton in silver,
and 0.40 ounce of gold. Near-surface
ores contained up to 12 ounces of silver, 0.95 ounce gold, 4 percent copper, and 5 percent lead.
Ore minerals: Arsenopyrite, pyrite, chalcopyrite, galena, sphalerite, jamesonite,
and realgar.

Production: 1890-1908, 300,000 tons of ore containing around 1,500,000 ounces of silver, and 120,000 ounces of gold.

References: Huntting, 1956, p. 162; Spurr, 1901, p. 803-804, p. 818-821; Landes and others, 1902, p. 71-72.

WHATCOM COUNTY

From 1900 to 1957, Whatcom County produced around 80,000 ounces of silver, most of which came from lode gold mines in the Mount Baker and Slate Creek mining districts. The county's major silver producers were the Great Excelsior, which produced around 65,000 ounces of silver from 1902 to 1915, and the Azurite gold mine, which produced 2,111 ounces of silver from 1936 to 1938. The remaining silver was chiefly a byproduct of gold mining operations at the Lone Jack, Red Mountain, New Light, and Mammoth mines. Only the Illinois mine in the Slate Creek district is classed as a silver mine because silver is the most valuable metal at this property. At other silver-bearing deposits in the county the value of gold, copper, lead, or zinc exceeds the value of silver.

With the exception of the Great Excelsior, silver-bearing deposits in the Mount Baker and Slate Creek districts are the quartz fissure type most of which are sparsely metallized. In general, galena-bearing veins carry the most silver, and up to several hundred ounces per ton in silver has been obtained from select samples. However, the average silver content of most veins is only around 5 to 10 ounces per ton. In a few veins galena, sphalerite, and chalcopyrite occur in small isolated ore shoots that contain up to several hundred tons of ore. At the Great Excelsior, silver occurs in pyritized and silicified breccia. The breccia averages around 3 ounces per ton in silver and 0.09 ounces per ton in gold. Although silver is present in the gold veins of the county, the average silver content of the veins is only from 0.02 to 0.15 ounce per ton. A high of 2.75 ounces per ton in silver was reported in high-

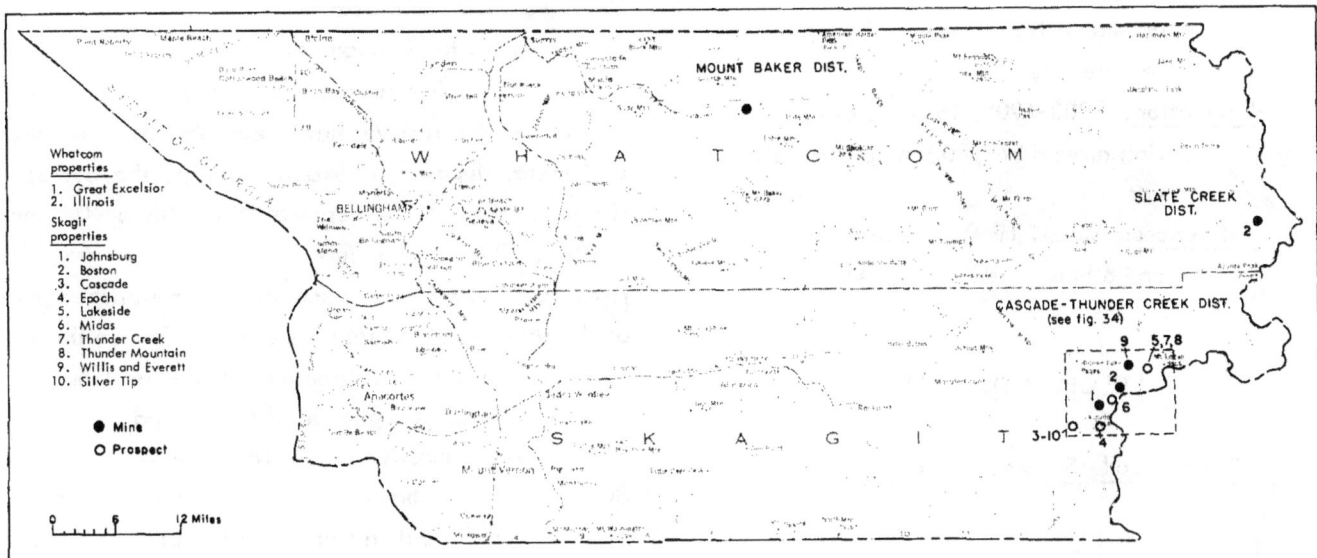

FIGURE 33.—Silver deposits of Whatcom and Skagit Counties.

grade gold ore from the Mammoth mine in the Slate
Creek district.

MOUNT BAKER DISTRICT

Principal Silver Deposit

Great Excelsior

Location: SW¼NW¼ sec. 6, T. 39 N.,
R. 8 E.

Development: 280-, 350-, and 450-foot
adit with almost 1,000 feet of drifts,
crosscuts, winzes, and raises. One
stope 180 feet long, 60 feet wide,
and up to 80 feet high.

Geology: Pyritized and silicified breccia in
andesite, argillite, and slate. Breccia
averages 0.09 oz. per ton in gold and
3.4 ozs. per ton in silver.

Ore minerals: Mainly pyrite with sparse chalcopyrite, galena, sphalerite, and
arsenopyrite; minor very fine-grained
tetrahedrite.

Production: 1903-1905, 1917; total production around $69,000 in gold and sil-
ver.

References: Moen, 1969, p. 86-87, Landes
and others, 1902, p. 43-44.

SLATE CREEK DISTRICT

Principal Silver Deposit

Illinois

Location: NE¼SE¼ sec. 35, T. 38 N., R. 17 E.

Development: 2 adits one of which is 400
feet long.

Geology: 3- to 6-foot-thick quartz vein in
slate and porphyry. Select samples
of vein assayed 18 to 600 ozs. silver,
and trace to 0.30 oz. gold.

Ore minerals: Galena, sphalerite, pyrargyrite,
tetrahedrite, free gold, and pyrite.

Production: Minor production in 1904.

Reference: Moen, 1969, p. 109-110.

SKAGIT COUNTY

The production of silver in Skagit County has
been insignificant; from 1900 to 1955, several small-
scale mining operations in the Cascade and Thunder
Creek districts of eastern Skagit County (fig. 33)
produced 2,613 ounces of silver. The bulk of the silver produced in the late 1890's and early 1900's came
from the Boston and Willis and Everett mines, near
the headwaters of Thunder Creek. Silver produced in
1953 and 1955 came from the Johnsburg mine near the
headwaters of the Cascade River.

Although assays of up to several hundred
ounces per ton in silver have been obtained from several metal deposits in Skagit County, the average
silver content of the deposits is less than 10 ounces
per ton. The silver occurs in shear zones in granodiorite, quartz diorite, and schist. The metallized parts
of the shear zones average less than 1 foot in thickness and contain disseminated sulfides or lenses of
massive sulfides in a gangue of quartz and calcite.
Some lenses of massive argentiferous galena are 4 to
30 inches thick; however, the lenses are seldom
greater than 50 feet in stope or pitch length. Common
sulfides of the veins include pyrite, arsenopyrite,
sphalerite, and chalcopyrite. The less common ore

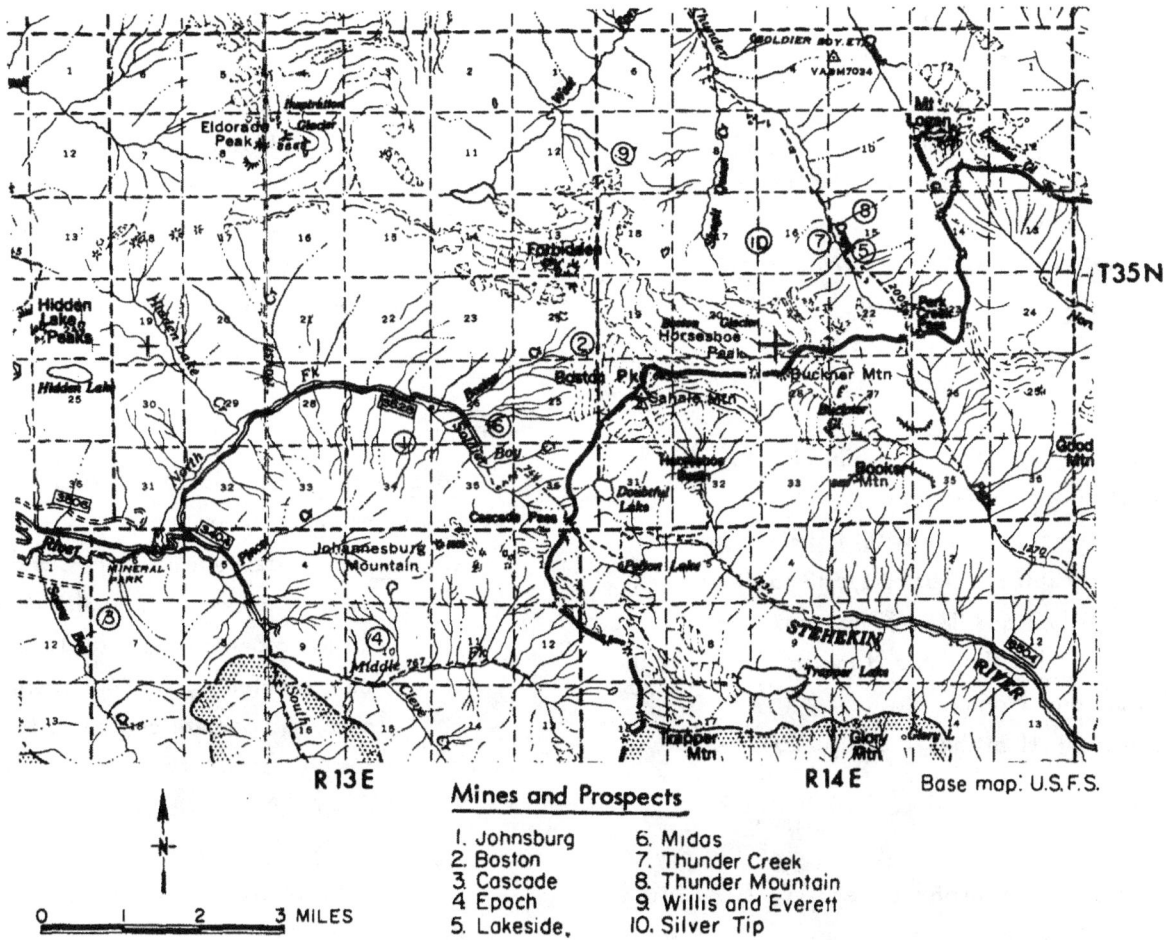

FIGURE 34.—Index map of the Cascade and Thunder Creek districts.

Mines and Prospects

1. Johnsburg
2. Boston
3. Cascade
4. Epoch
5. Lakeside.
6. Midas
7. Thunder Creek
8. Thunder Mountain
9. Willis and Everett
10. Silver Tip

minerals are cerussite, anglesite, and native gold. The silver-bearing minerals are argentiferous galena and fine-grained argentite.

The small size of the ore shoots, and their erratic distribution in the veins, appear to be main factors that discouraged mining operations in the past. Because most deposits fall within the boundaries of the North Cascades National Park, which was established in October 1968, it is unlikely that the deposits will ever be mined.

CASCADE AND THUNDER CREEK DISTRICTS

Principal Silver Deposits

Boston

Location: Sec. 24, T. 35 N., R. 13 E.

Development: 35- and 60-foot adits.

Geology: 6- to 9-foot-thick quartz vein in diorite. 8-inch band, which assayed

8.4 ozs. silver, 0.16 oz. gold, 0.2 percent copper, 22.1 percent lead, and 13 percent zinc.

Ore minerals: Galena, sphalerite, chalcanthite, chalcopyrite, pyrite, and arsenopyrite.

Production: 2 tons in the 1890's.

Reference: Huntting, 1956, p. 231.

Cascade

Location: NW¼ sec. 7, T. 34 N., R. 13 E.

Development: 100-foot adit.

Geology: 5-foot vein with a 10-inch band of sulfides that assays 48 to 51 ozs. silver and 32 to 37 percent lead.

Ore minerals: Galena and pyrite.

Production: None.

Reference: Huntting, 1956, p. 231.

Epoch

Location: Near center E½ sec. 10, T. 34 N., R. 13 E.

Development: Adit.

Geology: 3-foot vein of solid galena reported. Vein assays 39 to 102 ozs. silver and 38 to 45 percent lead.

Ore minerals: Galena.

Production: None.

Reference: Huntting, 1956, p. 231.

Johnsburg

Location: Secs. 27 and 34, T. 35 N., R. 13 E.

Development: 50- and 200-foot adits.

Geology: Metallized shear zone in schist. Ore minerals occur in lenses 4 to 30

inches thick and up to 50 feet long. Some ore assayed 57 percent lead and 40 ozs. per ton in silver.

Ore minerals: Galena, cerussite, anglesite, chalcopyrite, sphalerite, and pyrite.

Production: 1953 and 1955, total production 19 tons.

References: Huntting, 1956, p. 231.

Lakeside

Location: SE¼ sec. 15, T. 35 N., R. 14 E.

Development: Unknown.

Geology: Quartz veins in granite and metamorphic rocks. Some veins contain 23 to 66 percent galena, 16.32 percent sphalerite, 28.14 percent pyrite, and 2.58 percent chalcopyrite. Select samples contained 51.96 ozs. silver and 0.18 oz. gold.

Ore minerals: Galena, sphalerite, chalcopyrite, and pyrite.

Production: None.

Reference: Huntting, 1956, p. 232.

Midas

Location: W½ sec. 25 and E½ sec. 26, T. 35 N., R. 13 E.

Development: 2 adits with over 1,500 feet of drifts and crosscuts.

Geology: Narrow sulfide-bearing quartz veins in sheared diorite gneiss. Sulfides occur in lenses 7 to 24 inches in thickness and assay up to 20 ozs. silver, 0.14 oz. gold, 2.2 percent lead, and 1.96 percent zinc.

Ore minerals: Chalcopyrite, galena, sphalerite, and pyrite.

Production: None.

Reference: Huntting, 1956, p. 315, Geology
and Earth Resources files.

Silver Tip

Location: Secs. 16 and 17, T. 35 N.,
R. 14 E.

Development: 40-foot shaft, several short
adits, and open cuts.

Geology: 3- to 4-foot thick vein containing
1 to 2 feet of ore that assays 19.9 ozs.
silver, 20.5 percent lead, and 6.6
percent zinc.

Ore minerals: Galena, sphalerite, and
pyrite.

Production: None.

Reference: Huntting, 1956, p. 232.

Thunder Creek (Dorothy)

Location: W½ Sec. 15, T. 35 N., R. 14 E.

Development: 425- and 760-foot adits.

Geology: 6-inch to 6-foot-thick metallized
shear zone in altered diorite. Samples
from shear zone assayed 4.25 to 7 ozs.
silver, 5 to 10 percent lead, 4 to
4.7 percent zinc, and 0.34 percent
copper.

Ore minerals: Galena, chalcopyrite, sphalerite, and pyrite.

Production: None.

Reference: Huntting, 1956, p. 232.

Thunder Mountain

Location: Secs. 15, 22, and 29, T. 35 N.,
R. 14 E.

Development: Several short adits.

Geology: 1- to 6-foot-thick metallized
shear zones in gneiss and diorite.
Samples from shear zones assayed
6.82 to 29.34 ozs. silver, 0.30 to
4 percent copper, 0.4 to 18.7 percent
lead, and 0.4 to 14 percent zinc.
Select high-grade samples of galena
assayed 151 ozs. silver and 55.5
percent lead.

Ore minerals: Galena, sphalerite, chalcopyrite, pyrite, and pyrrhotite.

Production: Small test shipment.

Reference: Geology and Earth Resources files.

Willis and Everett

Location: Sec. 7, T. 35 N., R. 14 E.

Development: Unknown.

Geology: 3 veins, 4 to 12 feet thick in
granite and porphyry. Select high-
grade ore assayed 500 ozs. per ton in
silver.

Ore minerals: Unknown.

Production: Minor production in the 1890's.

Reference: Huntting, 1956, p. 316.

KING COUNTY

Silver produced in King County from 1903 to
1960 totaled 31,102 ounces valued at 25,120. Most
silver came from the Apex and Cleopatra mines in the
Miller River district of northeastern King County.
Silver was produced in the county as early as 1897,
but production figures are not available for years prior

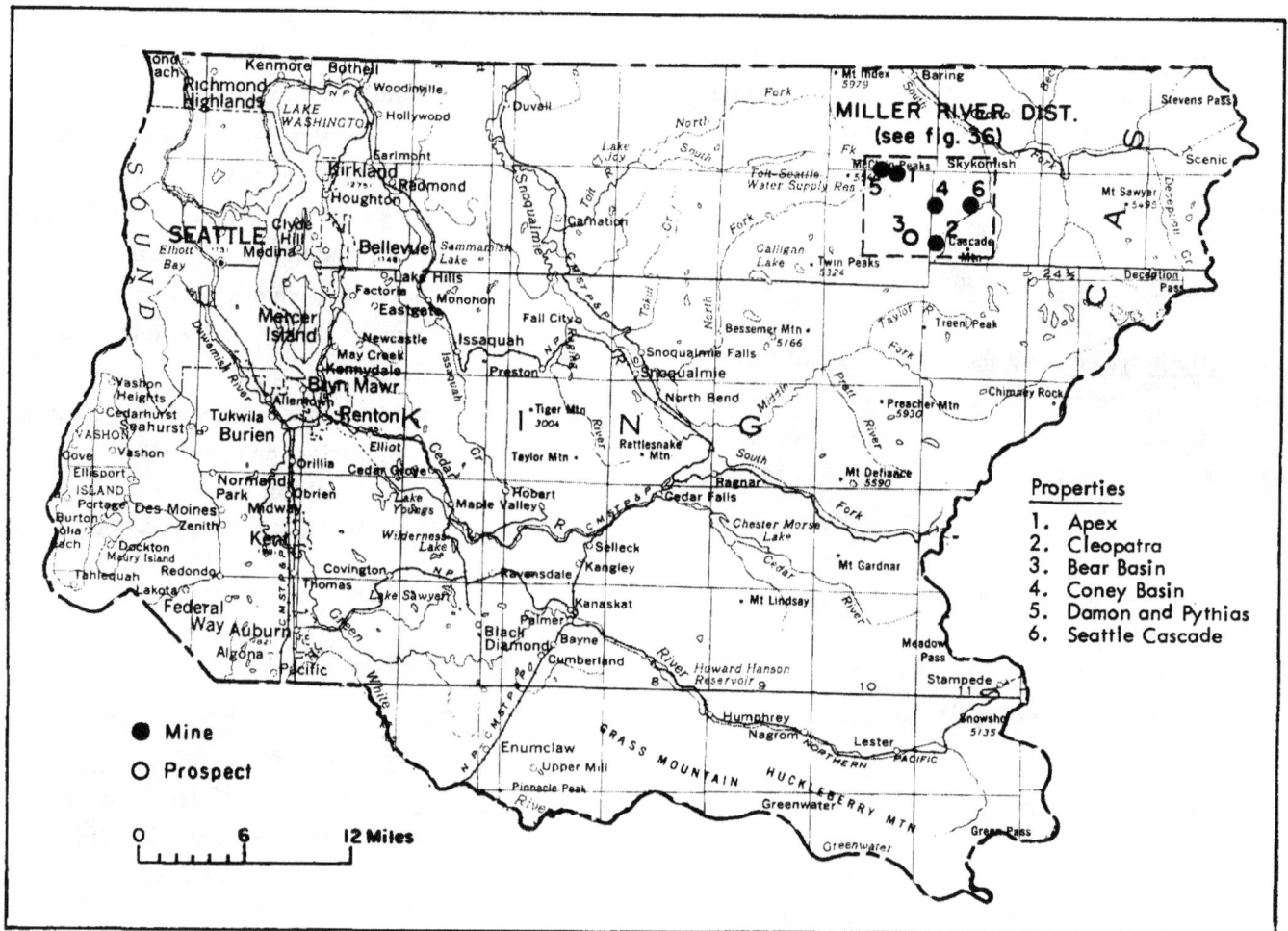

FIGURE 35.—Silver deposits of King County.

to 1904. The most productive silver mine was the Cleopatra, which produced around $250,000 in silver and gold. The bulk of the production came from near-surface silver ore that contained up to 250 ounces per ton in silver. Although the Apex is classed as a gold mine, the average silver content of ore mined was 9.15 ounces per ton. Total production of gold and silver from the Apex was around $300,000. Small shipments of copper, lead, zinc, antimony, gold, and silver ores were made from other mines in the county, but production from these mines was minor.

Mines and prospects that contain significant amounts of silver occur chiefly in the Miller River district. The ore minerals occur mainly in steeply dipping, northwest- to west-trending shear zones in granodiorite of the Snoqualmie batholith (middle Tertiary). Some shear zones contain metallized quartz-calcite veins, whereas other shear zones consist mainly of sheared and altered country rock with scattered grains of ore minerals. The common sulfides of the quartz-calcite veins and shear zones are pyrite, arsenopyrite, galena, sphalerite, and chalcopyrite; some deposits contain as much as 5 percent arsenic. The less common ore minerals are stibnite, molybdenite, tetrahedrite, jamesonite, stephanite, and stannite. Deposits that have high silver contents are also high in lead or gold.

Several deposits contain up to 100 ounces per ton in silver, but the ore shoots are generally less than 2 feet in thickness and are small. The ore minerals occur as narrow streaks in the veins and shear zones, or as scattered lenses and pods. In the Apex and Cleopatra mines, several ore shoots averaged 20 ounces per ton in silver and were as much as 200 feet long and up to 2½ feet thick.

MILLER RIVER DISTRICT

Principal Silver Deposits

Apex

Location: SW¼ sec. 34, T. 26 N., R. 10 E.

Development: 1,675-, 475-, 150-, and 100-foot adits over a vertical distance of 460 feet. Extensively stoped between the 1,675- and 475- foot adits.

Geology: Quartz vein 2 to 6 feet thick in granodiorite. High-grade ore occurs

Mines and Prospects

1. Apex
2. Cleopatra
3. Bear Basin
4. Coney Basin
5. Damon and Pythias
6. Seattle-Cascade

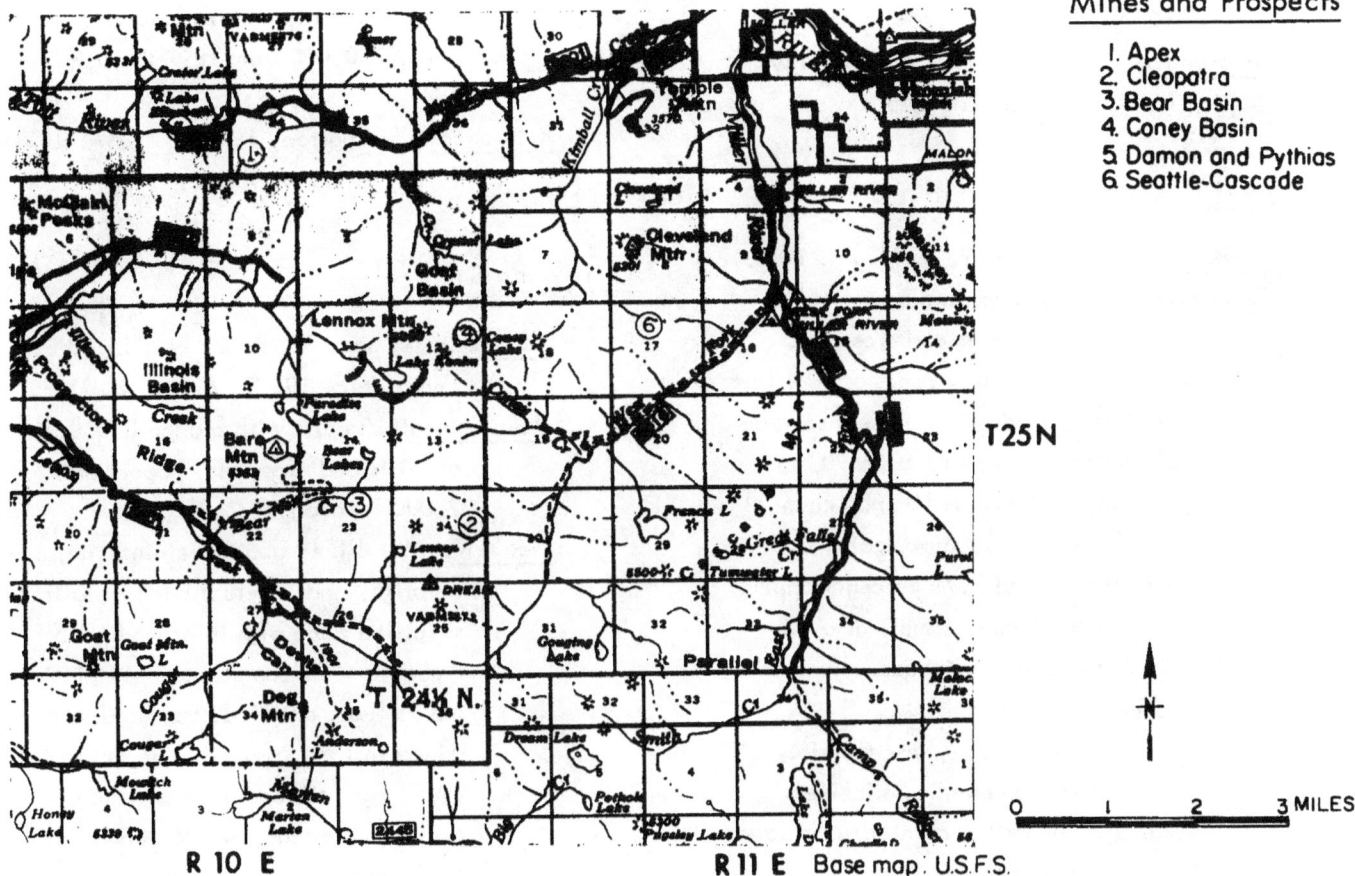

FIGURE 36.—Index map of the Miller River district.

as narrow streaks in the vein, and
assays up to 3.15 ounces per ton in
gold and 63 ounces per ton in silver.
Some ore contained 21 to 26 percent
arsenic, 10 to 20 ounces in silver,
$1\frac{1}{2}$ to $2\frac{1}{2}$ ounces in gold, and $4\frac{1}{2}$ to 6
percent lead. Average silver content
of 91 samples of vein in adits is 6.295
ounces per ton; average gold content
of samples is 1.125 ounces per ton.
Silver is contained in argentiferous
galena.

Ore minerals: Chalcopyrite, galena, sphalerite, arsenopyrite, and pyrite.

Production: $80,000 prior to 1901; 1905-1943,
$220,000.

References: Livingston, 1971, p. 145-147;
Patty, 1921, p. 301-305.

Bear Basin

Location: NE$\frac{1}{4}$ sec. 23, T. 25 N., R. 10 E.

Development: 8 adits containing a total of
2,165 feet of drifts and crosscuts.

Geology: Metallized shear zones up to 4 feet
thick in granodiorite. Select samples
of shear zones showed 10 to 90 ozs.
per ton of silver. A 10-inch channel
sample assayed 0.06 oz. gold, 6.0
ozs. silver, and 1.71 percent copper.
A 17-inch channel sample assayed
0.06 oz. gold, 16.90 ozs. silver, and
13 percent zinc.

Ore minerals: Minor pyrite, arsenopyrite,
freibergite, jamesonite, stibnite,
stannite, andorite, sphalerite, galena,
and molybdenite.

Production: None.

Reference: Livingston, 1971, p. 127-130.

Cleopatra

Location: NW$\frac{1}{4}$SE$\frac{1}{4}$ sec. 24, T. 25 N.,
R. 10 E.

Development: 1,260- and 160-foot adits
connected by raise.

Geology: 5-foot thick metallized shear zone
in granodiorite. Ore minerals are
spotty. One ore shoot $2\frac{1}{2}$ feet thick
and 100 feet long averaged 20 ounces
per ton in silver. Average silver content of 93 samples from vein in adits
is 13.35 ounces per ton.

Ore minerals: Argentiferous galena, chalcopyrite, tetrahedrite, sphalerite, jamesonite, pyrite, and arsenopyrite.

Production: $250,000 from 1897 through
1941.

References: Livingston, 1971, p. 139-140;
Purdy, 1951, p. 79-83.

Coney Basin

Location: N$\frac{1}{2}$ sec. 13, T. 25 N., R. 10 E.
and S$\frac{1}{2}$ sec. 19, T. 25 N., R. 11 E.

Development: 2 adits with around 3,000 feet
of drifts and crosscuts. Upper adit
2,000 feet long.

Geology: Metallized quartz veinlets in granodiorite, and a 4-foot wide metallized
silicified zone also in granodiorite.
Average of 22 samples showed 11.97
ozs. silver, and 0.38 oz. gold; ore
mined in 1941 assayed 0.86 oz. gold,
19.71 ozs. silver, 0.82 percent copper, 6.0 percent lead, 6 percent zinc,
1.52 percent arsenic, and 0.26 percent antimony.

Ore minerals: Galena, chalcopyrite, sphalerite, tetrahedrite, and pyrite.

Production: 40 tons in 1895, and minor production between 1934 and 1941.

Reference: Livingston, 1971, p. 140–141.

Damon and Pythias

Location: Center sec. 33, T. 26 N.,
R. 10 E.

Development: Several adits. Main adit is a 1,425-foot crosscut with 1,350 feet of drifts on two veins.

Geology: Two quartz veins up to 3 feet thick in granodiorite. Ore averages 0.245 oz. gold, 2.2 ozs. silver, and 7.86 percent arsenic. Ore shipped prior to 1940 averaged 0.87 oz. gold, 9 ozs. silver, and 4 percent lead.

Ore minerals: Chalcopyrite, galena, pyrite, and arsenopyrite.

Production: 23 tons prior to 1940.

Reference: Livingston, 1971, p. 147–148.

Seattle-Cascade

Location: Center N½ sec. 17, T. 25 N.,
R. 11 E.

Development: 400-foot adit, and an 800-foot adit with a 250-foot raise.

Geology: 1½-foot wide metallized shear zone in granodiorite. Shear zone contains sulfide bands up to 8 inches thick, which assay up to 30 ozs. silver per ton.

Ore minerals: Galena, sphalerite, chalcopyrite, pyrite, and arsenopyrite.

Production: Minor production prior to 1900, and in 1940.

Reference: Huntting, 1956, p. 297.

APPENDIX A

PRINCIPAL SILVER DEPOSITS OF EASTERN WASHINGTON BY COUNTY AND DISTRICT

CHELAN COUNTY
See Figure 22[1]/

Meadow Creek District

1. Iowa prospect
2. Orphan Boy prospect
3. Hunter prospect
4. Little Jap prospect
5. Sunday Morning mine

Stehekin District (Horseshoe Basin area)

6. Isoletta prospect
7. Doubtful prospect
8. Franklin prospect
9. Quien Sabe prospect
10. Homestake and Star prospect
11. Horseshoe Basin prospect

FERRY COUNTY
See Figure 13

Covada District

1. Gwin mine
2. Dan Patch prospect
3. Algonkian prospect

[1]/ Numbers on mines and prospects correspond to numbers on figures in text.

4. Longstreet mine
5. Silver Leaf mine
6. Shoo Fly mine
7. Silver Crown No. 2 mine
8. Ivanhoe prospect
9. Butterfly prospect
10. King Solomon prospect
11. Plymouth Rock prospect
12. Advance prospect
13. Keystone prospect
14. Royal Ann prospect
15. Good Ore prospect
16. Stray Dog mine
17. Oom Paul prospect
18. Montana mine
19. Romulus prospect
20. New York prospect
21. Meteor mine
22. Reserve prospect
23. Chief Barnaby prospect
24. Big Lake prospect
25. U.S. prospect

Keller District

26. McJunkin prospect
27. Congress prospect
28. Burlington and Delaware prospect
29. Shamrock mine
30. Anderson prospect
31. Summit prospect
32. Addison mine

33. Golden Chord mine
34. Jumper prospect
35. Rover Bonanza prospect
36. Meadow Creek prospect
37. Great Northern prospect

Curlew District

38. Anchor prospect
39. Panama prospect

KITTITAS COUNTY
See Figure 25

Cle Elum District

1. American Eagle prospect
2. Boss prospect
3. Aurora mine
4. Camp Creek mine
5. Cle Elum prospect

Gold Creek District

6. Silver King and Silver Queen prospect
7. Esther and Louisa mine
8. Transit prospect
9. Giant prospect

OKANOGAN COUNTY
See Figure 16

Loomis District

1. Mountain Sheep mine
2. Ruby mine
3. Number One prospect
4. Summit prospect
5. Eagle prospect
6. Kaaba-Texas mine

7. Chopaka prospect
8. Eureka prospect
9. Rich Bar prospect
10. Horn Silver mine
11. Bullfrog mine
12. Bellevue mine
13. Kimberly mine
14. Rainbow mine
15. Ivanhoe mine

Conconully District

16. Leuena mine
17. Mohawk prospect
18. Mammoth mine
19. Monitor mine
20. John Arthur prospect
21. Tough Nut mine
22. Lone Star mine
23. Key mine
24. Esther prospect
25. Salmon River prospect
26. Homestake mine
27. Lady of the Lake prospect
28. Leonora prospect
29. Nevada mine
30. Plant-Callahan mine
31. First Thought mine
32. Last Chance mine
33. Sonny Boy mine
34. Fourth of July mine
35. Keystone prospect
36. Arlington mine
37. Woo Loo Moo Loo prospect
38. Chloride prospect
39. Wind Fall prospect
40. Buck Mountain mine
41. Sherman mine
42. Silver Bell prospect

43. Silver Bluff mine
44. Black Huzzar prospect
45. Eureka prospect
46. Gold Eagle prospect
47. Gold Cup prospect
48. Lulu prospect
49. Evening Star prospect
50. Minnehaha prospect
51. Okanogan Belle prospect
52. Hardscrabble prospect
53. Central mine
54. Carl Fredrick prospect
55. Wheeler mine
56. Peacock mine
57. Shelby prospect

Sheridan District

58. Sheridan mine
59. Zalla M. mine
60. American Flag mine
61. Silver Bell mine

Park City District

62. Eureka prospect
63. Summit mine
64. Mountain Boy mine
65. Ramore prospect
66. Hanaford prospect

Nespelem District

67. Great Divide prospect
68. Anna prospect
69. Evening prospect
70. Andy O'Neil mine
71. Anchor prospect
72. Apache mine

73. Lilman mine
74. Little Chief mine
75. Panama mine

PEND OREILLE COUNTY
See Figure 7

Metaline District

1. Hanley mine
2. Oriole mine
3. Poorman mine
4. LaSota prospect
5. Rocky Creek mine

Newport District

6. Skippy and Queen Bess prospect
7. Eagle (Ries) mine

STEVENS COUNTY
See Figure 8

Orient District

1. Comstock prospect
2. Iron Mask prospect
3. Montana and Washington prospect
4. Easter Sunday mine
5. McKinley prospect

Northport District

6. Bullion mine
7. Great Republic prospect
8. Coyote mine
9. Sunset prospect
10. Sterret mine
11. Silver Crown mine
12. Melrose mine

13. Jackson mine
14. Red Top mine
15. Hazel mine
16. Roosevelt mine
17. United Treasure mine
18. Keough mine
19. Myeerah mine
20. Frisco Standard mine
21. Morning prospect
22. Burrus mine
23. Galena Farm prospect

Bossburg District

24. Al Ki prospect
25. Silver Trail mine
26. Young America mine
27. Bonanza mine
28. Gold Bar prospect
29. Chloride Queen mine
30. Avondale-Dome prospect

Kettle Falls District

31. Gold Ledge mine
32. Vanasse prospect
33. Gold Reef mine
34. Ark mine
35. Aguila mine

Colville District

36. Longshot mine
37. Middleport mine
38. Old Dominion mine
39. Ore Cache mine

Summit District

40. Daisy Tempest mine
41. Silver Summit mine

Chewelah District

42. Krug (Hanford) mine
43. Jay Dee mine
44. High Grade mine
45. United Copper mine
46. Copper King mine
47. Amazon mine
48. Chinto mine
49. Eagle (Blue Star) mine
50. Jay Gould mine
51. U.S. Copper Gold mine
52. Mullen prospect
53. Nevada mine
54. Edna mine
55. Double Eagle mine
56. Wells Fargo mine

Deer Trail District

57. Orchid mine
58. Cleveland mine
59. Deer Trail mine
60. Brooks mine
61. Saturday Night-Sunday Morning mine
62. Queen-Seal mine
63. Aichan Bee mine
64. Silver Star prospect
65. Orazada mine
66. Indian Trail mine

PRINCIPAL SILVER DEPOSITS OF WESTERN WASHINGTON
BY COUNTY AND DISTRICT

KING COUNTY
See Figure 35[1]

Miller River District

1. Apex mine
2. Cleopatra mine
3. Bear Basin prospect
4. Coney Basin mine
5. Damon and Pythias mine
6. Seattle Cascade mine

SKAGIT COUNTY
See Figure 33

Cascade-Thunder Creek District

1. Johnsburg mine
2. Boston mine
3. Cascade prospect
4. Epoch prospect
5. Lakeside prospect
6. Midas prospect
7. Thunder Creek prospect
8. Thunder Mountain prospect
9. Willis and Everett mine
10. Silver Tip prospect

SNOHOMISH COUNTY
See Figure 28

Darrington District

1. Larson prospect

[1] Numbers on mines and prospects correspond to numbers on figures in text.

2. Feldt prospect
3. Gray Mare prospect
4. Courtney prospect
5. Forest Hope prospect
6. Hunter prospect
7. Highland prospect
8. Hannah prospect
9. Green Crown prospect

Silverton District

10. New Seattle prospect
11. White Swan prospect
12. St. Louis and Jackson mine
13. Glengarry prospect
14. Big Four prospect

Sultan District

15. "45" (Magus) mine
16. Silver Horseshoe prospect
17. Calumet prospect

Silver Creek District

18. Jasper prospect
19. Jasperson prospect
20. Trade Dollar prospect
21. Mineral Center prospect
22. Morning Star prospect
23. Corona prospect
24. Crown Point prospect
25. Golconda prospect

26. Editor prospect
27. Billy Lee prospect
28. Blue Bird prospect
29. Vandalia prospect

Monte Cristo District

30. Monte Cristo (Mystery, Pride) prospect

WHATCOM COUNTY
See Figure 33

Mount Baker District

1. Great Excelsior mine

Slate Creek District

2. Illinois prospect

APPENDIX B

PROPERTY INDEX

Property	County	District	Index map page	Property description page
Addison	Ferry	Keller	99	100
Advance	Ferry	Covada	93	94
Aichan Bee	Stevens	Deer Trail	70	72
Algonkian	Ferry	Covada	93	94
Al Ki	Stevens	Bossburg	75	76
Amazon	Stevens	Chewelah	63	66
American Eagle	Kittitas	Cle Elum	141	141
American Flag	Okanogan	Sheridan	125	126
Anna	Okanogan	Nespelem	128	130
Anchor	Okanogan	Nespelem	128	130
Anchor	Ferry	Curlew	92	103
Anderson	Ferry	Keller	99	97
Andy O'Neil	Okanogan	Nespelem	128	130
Apache	Okanogan	Nespelem	128	130
Apex	King	Miller River	165	165
Aguila (Eagle-Newport)	Stevens	Kettle Falls	62	89
Ark (Silver Queen)	Stevens	Kettle Falls	62	89
Arlington	Okanogan	Conconully	106	110
Aurora	Kittitas	Cle Elum	141	141
Avondale-Dome	Stevens	Bossburg	75	76
Bear Basin	King	Miller River	163	166
Bellevue	Okanogan	Loomis	120	121
Big Four	Snohomish	Silverton	157	157
Big Lake	Ferry	Covada	92	94
Billy Lee	Snohomish	Silver Creek	153	152
Black Huzzar	Okanogan	Conconully	105	110
Blue Bird	Snohomish	Silver Creek	153	152
Buck Mountain	Okanogan	Conconully	106	110
Bonanza	Stevens	Bossburg	75	77
Boss	Kittitas	Cle Elum	141	142
Boston	Skagit	Cascade-Thunder Creek	161	161

Property	County	District	Index map page	Property description page
Brooks	Stevens	Deer Trail	70	72
Bullfrog	Okanogan	Loomis	120	122
Bullion	Stevens	Northport	62	85
Burlington and Delaware	Ferry	Keller	92	100
Burrus	Stevens	Northport	62	85
Butterfly	Ferry	Covada	93	94
Calumet	Snohomish	Sultan	151	151
Camp Creek	Kittitas	Cle Elum	141	142
Carl Frederick (Bernhardt)	Okanogan	Conconully	105	111
Cascade	Skagit	Cascade-Thunder Creek	161	162
Central	Okanogan	Conconully	105	111
Chief Barnaby	Ferry	Covada	92	95
Chinto	Stevens	Chewelah	63	66
Chloride	Okanogan	Conconully	106	111
Chloride Queen	Stevens	Bossburg	75	77
Chopaka	Okanogan	Loomis	120	122
Cle Elum	Kittitas	Cle Elum	141	142
Cleopatra	King	Miller River	165	166
Cleveland	Stevens	Deer Trail	62	72
Comstock	Stevens	Orient	62	91
Coney Basin	King	Miller River	165	166
Congress	Ferry	Keller	92	101
Copper King	Stevens	Chewelah	63	66
Corona	Snohomish	Silver Creek	153	152
Courtney	Snohomish	Darrington	155	155
Coyote	Stevens	Northport	115	85
Crown Point	Snohomish	Silver Creek	153	153
Daisy Tempest	Stevens	Summit	62	79
Damon and Pythias	King	Miller River	165	167
Dan Patch	Ferry	Covada	93	95
Deer Trail	Stevens	Deer Trail	70	73
Double Eagle	Stevens	Chewelah	62	66
Doubtful	Chelan	Stehekin	138	139

Property	County	District	Index map page	Property description page
Eagle	Okanogan	Loomis	120	122
Eagle (Blue Star)	Stevens	Chewelah	63	67
Easter Sunday	Stevens	Orient	62	91
Editor	Snohomish	Silver Creek	153	153
Edna	Stevens	Chewelah	62	67
Epoch	Skagit	Cascade-Thunder Creek	161	162
Esther	Okanogan	Conconully	106	111
Esther and Louisa	Kittitas	Gold Creek	142	142
Eureka	Okanogan	Conconully	105	111
Eureka	Okanogan	Park City	133	132
Eureka	Okanogan	Loomis	120	122
Evening	Okanogan	Nespelem	128	131
Evening Star	Okanogan	Conconully	106	111
Feldt	Snohomish	Darrington	155	156
First Thought	Okanogan	Conconully	106	112
Forest Hope	Snohomish	Darrington	155	156
Fourth of July	Okanogan	Conconully	106	112
"45" (Magus)	Snohomish	Sultan	151	152
Franklin	Chelan	Stehekin	138	139
Frisco Standard	Stevens	Northport	115	85
Galena Farm	Stevens	Northport	62	86
Giant	Kittitas	Gold Creek	142	143
Glengarry	Snohomish	Silverton	157	158
Golconda	Snohomish	Silver Creek	153	154
Gold Bar	Stevens	Bossburg	75	77
Gold Cup	Okanogan	Conconully	105	112
Gold Eagle	Okanogan	Conconully	105	112
Gold Ledge	Stevens	Kettle Falls	62	90
Gold Reef (Benevue)	Stevens	Kettle Falls	62	90
Golden Chord	Ferry	Keller	99	101
Good Ore	Ferry	Covada	93	95
Gray Mare	Snohomish	Darrington	155	156

Property	County	District	Index map page	Property description page
Great Divide	Okanogan	Nespelem	128	131
Great Excelsior	Whatcom	Mount Baker	159	160
Great Northern	Ferry	Keller	99	101
Great Republic	Stevens	Northport	115	86
Green Crown	Snohomish	Darrington	155	156
Gwin	Ferry	Covada	93	95
Hanford	Okanogan	Park City	133	133
Hanley	Pend Oreille	Metaline	58	59
Hannah	Snohomish	Darrington	155	156
Hardscrabble	Okanogan	Conconully	106	112
Hazel	Stevens	Northport	115	86
High Grade	Stevens	Chewelah	63	67
Highland	Snohomish	Darrington	155	156
Homestake	Okanogan	Conconully	106	113
Homestake and Star	Chelan	Stehekin	138	139
Horn Silver	Okanogan	Loomis	120	122
Horseshoe Basin	Chelan.	Stehekin	138	139
Hunter	Chelan	Meadow Creek	136	137
Hunter	Snohomish	Darrington	155	157
Illinois	Whatcom	Slate Creek	159	160
Indian Trail	Stevens	Deer Trail	62	73
Iowa	Chelan	Meadow Creek	136	137
Iron Mask	Stevens	Orient	62	91
Isoletta	Chelan	Stehekin	138	139
Ivanhoe	Ferry	Covada	93	95
Ivanhoe	Okanogan	Loomis	120	123
Jackson	Stevens	Northport	115	86
Jasper	Snohomish	Silver Creek	153	154
Jasperson	Snohomish	Silver Creek	153	154
Jay Dee	Stevens	Chewelah	63	67
Jay Gould	Stevens	Chewelah	63	67
John Arthur	Okanogan	Conconully	106	113

Property	County	District	Index map page	Property description page
Johnsburg	Skagit	Cascade-Thunder Creek	161	162
Jumper	Ferry	Keller	99	101
Kaaba-Texas	Okanogan	Loomis	120	123
Keough	Stevens	Northport	115	86
Key	Okanogan	Conconully	106	113
Keystone	Okanogan	Conconully	106	113
Keystone	Ferry	Covada	93	96
Kimberly	Okanogan	Loomis	120	123
King Solomon	Ferry	Covada	93	96
Krug (Hartford)	Stevens	Chewelah	62	68
Lady of the Lake	Okanogan	Conconully	106	113
Lakeside	Skagit	Cascade-Thunder Creek	161	162
Larson	Snohomish	Darrington	155	157
La Sota	Pend Oreille	Metaline	58	60
Last Chance	Okanogan	Conconully	106	114
Leonora	Okanogan	Conconully	106	114
Leuena	Okanogan	Conconully	106	114
Lilman	Okanogan	Nespelem	128	131
Little Chief	Okanogan	Nespelem	128	131
Little Jap	Chelan	Meadow Creek	136	137
Lone Star	Okanogan	Conconully	106	114
Longshot	Stevens	Colville	62	81
Longstreet	Ferry	Covada	93	96
Lulu	Okanogan	Conconully	105	114
Mammoth	Okanogan	Conconully	106	114
McJunkin	Ferry	Keller	92	101
McKinley	Stevens	Orient	62	91
Meadow Creek	Ferry	Keller	99	102
Melrose (Paragon)	Stevens	Northport	115	87
Meteor	Ferry	Covada	93	96
Midas	Skagit	Cascade-Thunder Creek	161	162
Middleport	Stevens	Colville	62	82

Property	County	District	Index map page	Property description page
Mineral Center	Snohomish	Silver Creek	153	154
Minnehaha	Okanogan	Conconully	106	115
Mohawk	Okanogan	Conconully	106	115
Monitor	Okanogan	Conconully	106	115
Montana	Ferry	Covada	93	96
Montana and Washington	Stevens	Orient	62	91
Monte Cristo	Snohomish	Monte Cristo	150	158
Morning	Stevens	Northport	62	87
Morning Star	Snohomish	Silver Creek	153	154
Mountain Boy	Okanogan	Park City	133	133
Mountain Sheep	Okanogan	Loomis	120	123
Mullen	Stevens	Chewelah	63	68
Myeerah	Stevens	Northport	115	87
Nevada	Stevens	Chewelah	62	68
Nevada	Okanogan	Conconully	106	115
New Seattle	Snohomish	Silverton	157	158
New York	Ferry	Covada	93	97
Number 1	Okanogan	Loomis	120	124
Okanogan Belle	Okanogan	Conconully	106	116
Old Dominion	Stevens	Colville	62	82
Oom Paul	Ferry	Covada	93	97
Orazada	Stevens	Deer Trail	70	73
Orchid	Stevens	Deer Trail	70	73
Ore Cache	Stevens	Colville	62	82
Oriole	Pend Oreille	Metaline	58	60
Orphan Boy	Chelan	Meadow Creek	136	137
Panama	Ferry	Curlew	92	103
Panama	Okanogan	Nespelem	128	132
Peacock	Okanogan	Conconully	106	116
Plant-Callahan	Okanogan	Conconully	106	116
Plymouth Rock	Ferry	Covada	93	97
Poorman	Pend Oreille	Metaline	58	60

Property	County	District	Index map page	Property description page
Quien Sabe	Chelan	Stehekin	138	140
Queen-Seal	Stevens	Deer Trail	70	74
Rainbow	Okanogan	Loomis	120	124
Ramore	Okanogan	Park City	133	134
Red Top	Stevens	Northport	115	87
Reserve	Ferry	Covada	93	97
Rich Bar	Okanogan	Loomis	120	124
Ries (Eagle)	Pend Oreille	Newport	58	60
Rocky Creek	Pend Oreille	Metaline	58	60
Romulus	Ferry	Covada	93	98
Roosevelt	Stevens	Northport	115	88
Rover Bonanza	Ferry	Keller	99	102
Royal Ann	Ferry	Covada	93	98
Ruby	Okanogan	Loomis	120	124
St. Louis and Jackson	Snohomish	Silverton	157	158
Salmon River	Okanogan	Conconully	106	116
Saturday Night-Sunday Morning	Stevens	Deer Trail	70	74
Seattle-Cascade	King	Miller River	165	167
Shamrock (Iron Creek)	Ferry	Keller	99	102
Shelby	Okanogan	Conconully	106	116
Sherman	Okanogan	Conconully	105	116
Sheridan	Okanogan	Sheridan	125	126
Shoo Fly	Ferry	Covada	93	98
Silver Bell	Okanogan	Sheridan	125	126
Silver Belle	Okanogan	Conconully	105	117
Silver Bluff	Okanogan	Conconully	105	117
Silver Crown	Stevens	Northport	115	88
Silver Crown No. 2	Ferry	Covada	93	97
Silver Horseshoe	Snohomish	Sultan	151	152
Silver King and Silver Queen	Kittitas	Gold Creek	142	143
Silver Leaf	Ferry	Covada	93	97
Silver Star	Stevens	Deer Trail	70	74
Silver Summit	Stevens	Summit	62	79

Property	County	District	Index map page	Property description page
Silver Tip	Skagit	Cascade–Thunder Creek	161	163
Silver Trail	Stevens	Bossburg	75	77
Skippy and Queen Bess	Pend Oreille	Newport	58	61
Sterrett	Stevens	Northport	115	88
Sonny Boy	Stevens	Conconully	106	117
Stray Dog	Ferry	Covada	93	98
Summit	Ferry	Keller	99	102
Summit	Okanogan	Park City	133	134
Summit	Okanogan	Loomis	120	124
Sunday Morning	Chelan	Meadow Creek	136	138
Sunrise	Okanogan	Conconully	. . .	117
Sunset	Stevens	Northport	115	88
Thunder Creek (Dorothy)	Skagit	Cascade–Thunder Creek	161	163
Thunder Mountain	Skagit	Cascade–Thunder Creek	161	163
Tough Nut	Okanogan	Conconully	106	117
Trade Dollar	Snohomish	Silver Creek	153	154
Transit	Kittitas	Gold Creek	142	143
United Copper	Stevens	Chewelah	63	68
United Treasure	Stevens	Northport	15	88
U.S.	Ferry	Covada	92	99
U.S. Copper Gold	Stevens	Chewelah	62	68
Vanasse	Stevens	Kettle Falls	62	90
Vandalia	Snohomish	Silver Creek	153	155
Wells Fargo	Stevens	Chewelah	62	69
Wheeler (Mineral Hill)	Okanogan	Conconully	106	118
White Swan	Snohomish	Silverton	157	158
Willis and Everett	Skagit	Cascade–Thunder Creek	161	163
Windfall	Okanogan	Conconully	106	118
Woo Loo Moo Loo	Okanogan	Conconully	106	118
Young America	Stevens	Bossburg	75	78
Zalla M.	Okanogan	Sheridan	125	127

APPENDIX C

SILVER PRODUCTION IN WASHINGTON

Year	Silver (troy ozs.)	Value (dollars)	Year	Silver (troy ozs.)	Value (dollars)
Prior to 1866 ...	120,500	161,108	1905	125,376	75,727
1866............	12,500	16,737	1906	45,878	30,738
1867............	5,000	6,650	1907	55,359	36,357
1868............	5,000	6,650	1908	88,823	47,076
1869............	3,750	4,969	1909	79,488	41,334
1870............	3,750	4,969	1910	205,345	110,886
1871............	4,000	5,300	1911	243,781	129,204
1872............	3,250	4,297	1912	413,538	254,326
1873............	2,355	3,054	1913	331,239	200,068
1874............	1,925	2,460	1914	264,861	146,468
1875............	1,000	1,240	1915	255,837	129,709
1876............	320	371	1916	355,121	220,510
1877............	41,667	50,000	1917	282,320	232,632
1878............	21,739	25,000	1918	310,093	310,093
1879............	17,857	20,000	1919	259,384	250,510
1880............	886	1,019	1920	199,678	217,649
1881............	1,450	1,639	1921	142,450	142,450
1882............	1,450	1,639	1922	205,046	205,046
1883............	387	430	1923	227,187	186,293
1884............	912	1,012	1924	213,742	143,207
1885............	52,208	55,863	1925	166,425	115,499
1886............	123,752	122,514	1926	171,649	107,109
1887............	94,516	92,626	1927	155,850	88,367
1888............	123,750	116,325	1928	99,738	58,347
1889............	81,984	77,035	1929	47,182	25,148
1890............	69,628	73,109	1930	32,816	12,634
1891............	165,883	164,224	1931	22,410	6,499
1892............	151,554	131,852	1932	17,412	4,910
1893............	134,961	105,270	1933	18,520	6,482
1894............	9,683	6,100	1934	44,120	28,522
1895	109,060	70,889	1935............	52,338	37,618
1896	233,407	158,717	1936............	66,900	51,814
1897	242,780	145,668	1937............	126,304	97,696
1898	329,549	194,434	1938............	380,938	246,263
1899	289,661	173,796	1939............	442,063	300,067
1900	302,570	187,593	1940	365,175	259,680
1901	377,381	222,466	1941	402,030	285,888
1902	721,450	376,308	1942	369,038	262,427
1903	305,401	201,789	1943	370,440	263,424
1904	157,598	89,831	1944	321,608	228,699

Year	Silver (troy ozs.)	Value (dollars)	Year	Silver (troy ozs.)	Value (dollars)
1945	281,444	200,138	1955	436,348	394,917
1946	264,453	213,678	1956 ... 1/ ...	448,441	405,862
1947	293,736	265,831	1957–1972 1/	6,410,835	7,822,417
1948	375,831	340,146			
1949	357,853	328,875	Total	22,822,490	$20,307,881
1950	363,656	329,127			
1951	335,000	303,145			
1952	316,000	285,675			
1953	321,202	290,704			
1954	313,715	283,928			

1/ Individual years were not published by U.S. Bureau of Mines so individual company confidential information would not be disclosed.

SELECTED REFERENCES

Addie, G. G., 1970, The Metaline district, Pend Oreille County, Washington. In Weissenborn, A. E.; Armstrong, F. C.; Fyles, J. T., editors, Lead-zinc deposits in the Kootenay Arc, northeastern Washington and adjacent British Columbia: Washington Division of Mines and Geology Bulletin 61, p. 65-78.

Bancroft, Howland, 1914, The ore deposits of northeastern Washington: U.S. Geological Survey Bulletin 550, 215 p.

Bethune, G. A., 1891, Mines and minerals of Washington: Washington State Geologist, First Annual Report, 122 p.

Bethune, G. A., 1892, Mines and minerals of Washington: Washington State Geologist, Second Annual Report, 1891, 187 p.

Broughton, W. A., 1942, Inventory of mineral properties in Snohomish County, Washington: Washington Division of Geology Report of Investigations 6, 64 p.

Campbell, A. B.; Raup, O. B., 1964, Preliminary geologic map of the Hunters quadrangle, Stevens and Ferry Counties, Washington: U.S. Geological Survey Mineral Investigations Field Studies Map MF-276, scale 1:48,000.

Carithers, Ward; Guard, A. K., 1945, Geology and ore deposits of the Sultan Basin, Snohomish County, Washington: Washington Division of Mines and Geology Bulletin 36, 90 p.

Clark, L. D.; Miller, F. K., 1968, Preliminary geologic map of the Chewelah Mountain quadrangle, Stevens County, Washington: Washington Division of Mines and Geology Geologic Map GM-5, scale 1:62,500.

Colville Engineering Company, 1941, Report on minerals in Pend Oreille County: Preliminary draft (assembled for the Public Utility District of Pend Oreille County, Washington, by the Colville Engineering Company), 75 p.

Colville Engineering Company, 1941, Report on minerals in Stevens County: Preliminary draft (assembled for the Public Utility District of Stevens County, Washington, by the Colville Engineering Company), 137 p.

Culver, H. E.; Broughton, W. A., 1945, Tungsten resources of Washington: Washington Division of Geology Bulletin 34, 89 p.

Fox, Kenneth F., Jr.; Rinehart, C. Dean, 1972, Distribution of copper and other metals in gully sediments of part of Okanogan County, Washington: Washington Division of Mines and Geology Bulletin 65, 38 p.

Fulkerson, F. B.; Kingston, G. A., 1958, Mine production of gold, silver, copper, lead, and zinc in Pend Oreille and Stevens Counties, Washington, 1902-1956: U.S. Bureau of Mine Information Circular 7872, 51 p.

Full, Roy P.; Grantham, Robert M., 1968, Ore deposits of the Republic mining district, Ferry County, Washington. In Ridge, John D., editor. Ore deposits of the United States, 1933-1967—The Graton-Sales Volume: American Institute of Mining, Metallurgic, and Petroleum Engineers, New York City, volume 2, p. 1481-1494.

Gage, H. L., 1941, The zinc-lead mines of Washington: U.S. Bonneville Power Administration, Market Development Section, 235 p.

SELECTED REFERENCES—Continued

Goldsmith, Richard, 1952, Petrology of the Tiffany-Conconully area, Okanogan County, Washington: University of Washington Ph.D. thesis, 356 p.

Gualtieri, J. L.; Simmons, G. C.; Thurber, H. K.; Miller, M. S., 1973, Mineral resources of the Alpine Lakes study area, Chelan, King, and Kittitas Counties, Washington: U.S. Geological Survey open-file report, 132 p.

Hodges, L. K., editor, 1897, Mining in the Pacific Northwest: The Seattle Post-Intelligencer, Seattle, Washington, 192 p. Fascimite Reproduction SJS-134 by the Shorey Book Store, Seattle, Washington in 1967.

Hundhausen, R. J., 1949, Investigation of the Young America lead-zinc deposit, Stevens County, Washington: U.S. Bureau of Mines Report of Investigations 4556, 13 p.

Huntting, M. T., 1943, Inventory of mineral properties in Chelan County, Washington: Washington Division of Geology Report of Investigations 9, 63 p.

Huntting, M. T., 1956, Inventory of Washington minerals—Part 2, Metallic minerals: Washington Division of Mines and Geology Bulletin 37, v. 1, 428 p., v. 2, 67 p.

Jenkins, O. P., 1924, Lead deposits of Pend Oreille and Stevens Counties, Washington: Washington Division of Geology Bulletin 31, 153 p.

Jones, E. L., Jr., 1916, Reconnaissance of the Conconully and Ruby mining districts, Washington: U.S. Geological Survey Bulletin 640-B, p. 11-36.

Landes, Henry; Thyng, W. S.; Lyon, D. A.; Roberts, Milnor, 1902, The metalliferous resources of Washington, except iron: Washington Geological Survey Annual Report for 1901, part 2, 123 p.

Livingston, V. E., Jr., 1971, Geology and mineral resources of King County, Washington: Washington Division of Mines and Geology Bulletin 63, 200 p.

Magill, E. A.; Schlagel, J. C., 1962, Copper deposits in the Silver Creek mining district, Snohomish County, Washington: U.S. Bureau of Mines open-file report, 59 p.

Menzer, F. J., 1964, Geology of the crystalline rocks west of Okanogan, Washington: University of Washington Ph.D. thesis, 64 p.

Moen, W. S., 1969, Mines and mineral deposits of Whatcom County, Washington: Washington Division of Mines and Geology Bulletin 57, 134 p.

Moen, W. S., 1973, Conconully mining district of Okanogan County, Washington: Washington Division of Mines and Geology Information Circular 49, 42 p.

Muessig, Siegfried, 1967, Geology of the Republic quadrangle and a part of the Aeneas quadrangle, Ferry County, Washington: U.S. Geological Survey Bulletin 1216, 135 p.

Pardee, J. T., 1918, Geology and mineral deposits of the Colville Indian Reservation, Washington: U.S. Geological Survey Bulletin 677, 186 p.

Park, C. F., Jr.; Cannon, R. S., Jr. 1943, Geology and ore deposits of the Metaline quadrangle, Washington: U.S. Geological Survey Professional Paper 202, 81 p.

Patty, E. N., 1921, The metal mines of Washington: Washington Geological Survey Bulletin 23, 366 p.

SELECTED REFERENCES—Continued

Pearson, R. C., 1967, Geologic map of the Bodie Mountain quadrangle, Ferry and Okanogan Counties, Washington: U.S. Geological Survey Geological Quadrangle Map GQ-636, scale 1:62,500.

Poole, A. A., 1939, The Deer Trail mine, Stevens County, Washington: University of Idaho School of Mines B.S. thesis, 10 p.

Purdy, C. P., Jr., 1951, Antimony occurrences of Washington: Washington Division of Mines and Geology Bulletin 39, 186 p.

Rinehart, C. D.; Fox, K. F., 1972, Geology and mineral deposits of the Loomis quadrangle, Okanogan County, Washington: Washington Division of Mines and Geology Bulletin 64, 124 p.

Fox, Kenneth F., Jr.; Rinehart, C. Dean, 1972, Distribution of copper and other metals in gully sediments of part of Okanogan County, Washington: Washington Division of Mines and Geology Bulletin 65, 38 p.

Schroeder, M. C., 1952, Geology of the Bead Lake district, Pend Oreille County, Washington: Washington Division of Mines and Geology Bulletin 40, 57 p.

Spurr, J. E., 1901, The ore deposits of Monte Cristo, Washington: U.S. Geological Survey 22nd Annual Report, part 2, p. 777-865.

Staatz, M. H., 1964, Geology of the Bald Knob quadrangle, Ferry and Okanogan Counties, Washington: U.S. Geological Survey Bulletin 1161-F, 79 p.

Thomson, J. P., 1973, The Swauk mining district: The Pacific Northwesterner, v. 17, no. 3, Summer 1973, p. 37-50.

Trimble, W. J., 1914, The mining advance into the Inland Empire: University of Wisconsin Ph.D. thesis, 255 p.

Umpleby, J. B., 1910, Geology and ore deposits of Republic mining district: Washington Geological Survey Bulletin 1, 65 p.

Umpleby, J. B., 1911, Geology and ore deposits of the Oroville-Nighthawk mining district: Washington Geological Survey Bulletin 5, part 2, p. 53-107, 110-111.

U.S. Bureau of Mines, 1928 to date, Minerals Yearbook.

U.S. Geological Survey, 1883 to 1927, Mineral Resources of the United States.

Weaver, C. E., 1911, Geology and ore deposits of the Blewett mining district: Washington Geological Survey Bulletin 6, 104 p.

Weaver, C. E., 1913, Geology and ore deposits of the Covada mining district: Washington Geological Survey Bulletin 16, 87 p.

Weaver, C. E., 1920, The mineral resources of Stevens County: Washington Geological Survey Bulletin 20, 350 p.

Webb, R. T., 1957, Petrology, structure, and mineralization of the Meadow Creek area, Chelan County, Washington: University of Arizona M.S. thesis, 51 p.

SELECTED REFERENCES—Continued

Weissenborn, A. E.; Armstrong, F. C.; Fyles, J. T., 1970, Lead-zinc deposits in the Kootenay Arc, northeastern Washington and adjacent British Columbia: Washington Division of Mines and Geology Bulletin 61, 123 p.

Western Historical Publishing Company, 1904, An illustrated history of Stevens, Ferry, Okanogan, and Chelan Counties, State of Washington: Western Historical Publishing Company, Spokane, 867 p.

Yates, R. G., 1970, Geologic background of the Metaline and Northport mining districts, Washington. In Weissenborn, A. E.; Armstrong, F. C.; Fyles, J. T., editors, Lead-zinc deposits in the Kootenay Arc, northeastern Washington and adjacent British Columbia: Washington Division of Mines and Geology Bulletin 61, p. 17-39.

www.ingramcontent.com/pod-product-compliance
Lightning Source LLC
Chambersburg PA
CBHW051213200326
41519CB00025B/7096